Science and
Ideology in
Soviet Society
1917-1967

Science and Ideology in Soviet Society

1917-1967

George Fischer
Richard T. De George
Loren R. Graham
Herbert S. Levine
Edited by George Fischer

Routledge
Taylor & Francis Group

LONDON AND NEW YORK

Originally published in 1967 by Atherton Press.

Published 2012 by Transaction Publishers

Published 2017 by Routledge
2 Park Square, Milton Park, Abingdon, Oxon OX14 4RN
711 Third Avenue, New York, NY 10017, USA

Routledge is an imprint of the Taylor & Francis Group, an informa business

Library of Congress Catalog Number: 2011035590

Library of Congress Cataloging-in-Publication Data

Science & ideology in Soviet society
    Science and ideology in Soviet society : 1917-1967 / George Fischer ...
    [et al.].
        p. cm.
    Originally published: New York : Atherton Press, 1967, with title Science
    & ideology in Soviet society.
    ISBN 978-1-4128-4594-6 (acid-free paper)
        1. Learning and scholarship--Soviet Union--History. 2. Science--So-
    viet Union--History. 3. Science--Political aspects--Soviet Union--History.
    4. Science--Social aspects--Soviet Union--History. 5. Ideology--Soviet
    Union--History. 6. Soviet Union--Intellectual life--1917-1970. 7. Soviet
    Union--Politics and government--1917-1936. 8. Soviet Union--Politics
    and government--1936-1953. 9. Soviet Union--Politics and government-
    -1953-1985. I. Fischer, George, 1923- II. Title.
AZ712.S3 2012
001.20947--dc23

                                                        2011035590

ISBN 13: 978-1-4128-4594-6 (pbk)

THE AUTHORS

GEORGE FISCHER, who is also the editor of this book, is an Associate Professor of Sociology at Columbia University and, concurrently, a staff member of the Russian Institute and the Bureau of Applied Social Research. Professor Fischer has held research appointments as a Junior Fellow of the Society of Fellows, Harvard University, as a Fellow of the Center for Advanced Study in the Behavioral Sciences, and as a Guggenheim Fellow. He has written five and edited three earlier books in sociology and history. His next book, *The Soviet System and Modern Society*, will also be published by Atherton Press.

RICHARD T. DE GEORGE was awarded his Ph.D. at Yale and is now Professor and Chairman of the Department of Philosophy at the University of Kansas. From 1965 to 1966 he was a Lecturer in Philosophy and a Senior Research Fellow at the Russian Institute of Columbia University. Professor De George has held Fulbright, Ford Foundation, and ACLS/SSRC fellowships. He has edited and contributed to several books, and is the author of *Patterns of Soviet Thought*.

LOREN R. GRAHAM is an Associate Professor and a staff member of the Russian Institute at Columbia University. Professor Graham began his career as a research engineer. Specializing in the history of science, he was awarded his Ph.D. in history at Columbia. He taught at Indiana University before returning to Columbia. He is the author of *The Soviet Academy of Sciences and the Communist Party, 1927–1932*.

HERBERT S. LEVINE is an Associate Professor at the University of Pennsylvania and was a visiting professor of economics at the Russian Institute from 1964 to 1966. While a research fellow at the Russian Research Center of Harvard University, Professor Levine completed a large-scale study of economic planning in Soviet industry. He has contributed extensively to economic journals and has served as Executive Secretary of the Association for the Study of Soviet-type Economies (ASTE).

PREFACE

Our era has been called the age of science but
also the age of ideologies. Like most other societies today,
the Soviet Union responds to the impact of both science
and ideology, and there both elements loom especially
large. What is the interaction of the two? Laymen as
much as experts are apt to disagree sharply about, or be
ignorant of, their mutual effect.

This volume seeks to shed some light on this important
but difficult question. It focuses on the current Soviet
scene: each of the four chapters is devoted to a different
field of science—sociology, philosophy, cybernetics, and

economics. Recent developments have made all these fields of particular interest. Without claiming that these four areas are typical of all science or all fields of scholarship, the authors believe that the disciplines discussed are revealing of trends in Soviet science in general and of its interaction with an established (though not immutable) ideology in particular.

These essays grew out of a series of lectures at the Russian Institute of Columbia University delivered by four of the Institute's associates in 1966. The opening essay has been considerably revised and enlarged; this was done in part so as to provide some fuller introduction to the over-all problem of science and ideology. In its present version, then, it serves also as an introduction to the volume. The other essays, reworked by their authors, follow by and large the scope of the original lectures.

Each author deals with the interaction of science and ideology in a distinct way. The reader will find that, as he moves from the first essay to the last, the amount of attention explicitly paid to this problem diminishes. Instead, he will find increasing stress put on specialized issues within each field. This is especially true of the chapter on economics.

Nonetheless, there is a common theme running through the essays. While each author weighs and interprets the evidence quite distinctively, all lean toward one common view. They appear to agree that, in its present Soviet setting, the encounter of science with ideology need not always result in conflict. Specifically, they hold—in varying degrees—that in the 1960s Soviet science (or at least the disciplines covered here) may in fact help to sustain the established system and its ideology rather than weaken or erode them.

Without necessarily considering this a lasting pattern, each author finds that in his own field the evidence fails to bear out the opposite view, so widely held in the West. In contrast with many writers and observers who speak of an endemic and ubiquitous conflict between science and Soviet ideology, the contributors to this volume point to some interesting elements of harmony or even mutual reinforcement.

The authors are most grateful to the four scholars who kindly agreed to chair the sessions at which the original lectures were given: Professors Daniel Bell, Robert K. Merton, and William S. Vickrey, of Columbia University, and Sir Isaiah Berlin, of Wolfson College, Oxford. By their comments they contributed a good deal to whatever value this book may have.

The authors also wish to record their debt and gratitude to three members of the Russian Institute staff: Alexander Dallin, director until 1967; Lyn Cannastra, administrative assistant; and Constance A. Bezer, editorial assistant. Over the past few years, they have added much to the authors' various associations with the Institute. As it marked its twentieth year, the Russian Institute itself made possible this new look at a thorny problem that in our time affects all societies.

GEORGE FISCHER

CONTENTS

Preface

SOCIOLOGY

GEORGE FISCHER

Today, all agree, confusion and polemics mark scholarly discussion of ideology. This is as true of writings on one society, like the Soviet Union, as it is of any other.

Some sociologists thus view ideology pejoratively. They see it as a means by which interest groups seek to justify their position in the existing order or to change that order for their own benefit. Others, however, view ideology a lot less negatively. In that case, they see it as the general pattern of beliefs and values of any group. Such a cultural pattern makes action possible by providing order and meaning in problematic situations. Another clash stands out, too: while some writers identify ideology with the central value system of a society, others set up a dichotomy by which ideology is counterposed to the central value system. Each choice, as usual, leaves out things that the other does not.

We can get away from these controversies by conceiving of an ideological *dimension*, rather than of ideology as a distinct mode of thought. Such a dimension will be found in *any* group's system of beliefs, norms, and values that have to do with social action. The essence of this dimension lies in its main function: to legitimate social action.

The ideological dimension moves to the fore when established justifications become so problematic that a group feels forced to do two things to its tacit legitimating ideas: to make them more explicit, and to add new ideas or modify the old. Just how does a group shift back and forth between institutionalized and problematic legitimation? That depends on a great many variables, of which we have no clear picture as yet.

In this context, the classic problem of science and ideology assumes special importance for any science of society. Scholars in the social sciences must take ideology into account in their work; it affects the social action of the people they study and their own action as well. Ever since Marx raised it, the problem has been the same: What is the tie between science and ideology? On the surface the two seem to have little in common. Yet when we look at the contrasts more closely, the two turn out to be much harder to set apart.

1. Ideology tends to be thought of as a scheme of social action, and science as a scheme of pure knowledge. Science, in this view, aims at theory; ideology at practice. A group's ideology tells the group and others how to treat a social system; science seeks knowledge for its own sake.

However, both science and ideology can be viewed as schemes of social action. Science, as well as ideology, is concerned with *both* theory and practice. The rules of science also tell scholars what their orientation to the social system should be. Sometimes the rule holds that the nature and goals of social system are irrelevant to their work. At other times the rule is that scientific in vestigation should be concerned with problems that are socially useful and important. Beyond that, scholars are often not concerned with knowledge for its own sake, even when their rule of science tells them that they should be. Instead, they may not only aim at knowledge for the sake of a client or the powers that be but also produce only that type of knowledge.

2. In times of change, both science and ideology can serve as schemes that bridge the gap between reality and prior values and beliefs. According to the typical view, though, science does this by and large in specialized areas and for experts only. Ideology, it is held, is more general in scope and appeals to a wider public. And it makes more use of untestable claims about reality.

One may doubt, though, whether this is the case. Belief systems and groups attached to them which are called "ideological" assert highly esoteric and exclusive knowledge which nonmembers are said to lack. Thus scholars are not the only ones to rest their case on specialization and expertise. It is questionable, too, whether ideology makes much more use of untestable claims about reality. Any mode of scientific investigation is necessarily based on a whole set of assumptions about reality that cannot be tested.

3. Typically, ideology is seen as standing close to the "sacred" and as eliciting a good deal of affect. Science, it is argued, shows less of either. Yet the norms and goals of science can also be seen as "sacred." As Thomas Kuhn

points out, moreover, in times of crisis in science debates among scholars can show a great deal of affect.

The distinctions between science and ideology are, then, by no means clear. The classic problem is still very much with us, as we can see in the field of sociology. Clifford Geertz has put it well: "Where, if anywhere, ideology leaves off and science begins has been the Sphinx's Riddle of much of modern sociological thought and the rustless weapon of its enemies."[1] The Sphinx's Riddle takes a special form in the "new sociology" that has risen in the Soviet Union since Stalin. In its own way, the Soviet case brings out the difficulty of drawing a line between science and ideology.

In the West, it has been widely assumed that Soviet ideology runs counter to science and must clash with it. According to this view, a communist type of society necessarily leads to conflict between science and ideology, whereas in the West harmony between them can and often does exist. Applied to any field like sociology, this view entails an all-too-simple forecast: either ideology prevails, or science does. There is no third way, as Soviet writers like to say; *tertium non datur.*[2] I do not share this view. In the Soviet "new sociology," I think, more harmony may in fact exist between science and ideology than the Western view has it.

Whatever their disagreements, Western writers on the subject tend to concur on some matters. In communist-ruled societies, "ideology" refers to a *specific* body of ideas, an explicit and cohesive one. These ideas are *established*, or given dominance, by the society's ruling group. This situation affects deeply the way scientists speak of social reality. They do so in ideological terms. To the extent that science and ideology reflect opposing cultural or symbolic styles, Soviet scientists follow the ideological style when they speak of human relations.

Hence it is apparent that we cannot apply to Soviet students of society the dichotomy which Geertz set up between the styles of science and ideology:

> Science names the structure of situations in such a way that the attitude contained toward them is one of disinterestedness. Its style is restrained, spare, resolutely analytic: by shunning the semantic devices that most effectively formulate moral sentiment, it seeks to maximize intellectual clarity. But ideology names the structure of situations in such a way that the attitude contained toward them is one of commitment. Its style is ornate, vivid, deliberately suggestive: by objectifying moral sentiment through the same devices that science shuns, it seeks to motivate action.[3]

What is in fact new in the post-Stalin "new sociology"? The answer seems fairly plain. In limited areas and to a limited extent its cultural style has begun to change from a more "ideological," as Geertz speaks of it, to a more "scientific" one. This does not modify much the official tie of science to the state (and to its ideology). Yet the data now point to a revealing shift in the ways in which science and ideology relate to each other.

In rough outline at least, the early history of the "new sociology" in the Soviet Union is well known.[4] Although it hardly existed under Stalin, academic research on social problems became the object of official concern at the turn of the 1960s. Groups of Soviet scholars began to take part in international congresses of sociology. In the field of Marxist social philosophy, a few young scholars made plans to add "concrete" quantitative research to the deductive Soviet forms of social analysis. And scholars as well as officials argued with each other about the proper role of the "new sociology" in the context of Marxism-Leninism.

This chapter does not deal with the years when Soviet

sociology was embryonic; our focus is on the years since the early 'sixties. We now have some evidence about the actual "first steps" of the new sociology. In addition to a large amount of lesser writing, these first steps consist of a dozen or so books by Soviet sociologists from the early 'sixties up to the end of 1966. I shall refer to these books in sketching the relations between science and ideology typical of current Soviet work.[5] In this work, a few main trends are evident:

1. In recent years an official decision has been made to define "sociology" in terms not of philosophy but of specialized social research.

2. A "future" orientation treats doctrine about the future as the main guide to analyzing the present.

3. A second, "concrete" orientation increasingly emphasizes quantitative social research.

4. Western sociology gets much attention, especially criticisms.

5. On the level of theory, Soviet sociologists may now tackle a major problem raised by Marx—that of work and alienation.

This chapter outlines each of these trends.[*]

Philosophy or Social Research?

For the Soviet system, and for the studies of society it favors, one fact is basic. "Science" has two very different meanings. In the Soviet Union as elsewhere, the term refers to rigorous and verifiable methods of seeking knowledge. Equally important is a second official usage; science is wholly equated with established ideology, the

---

[*] This chapter is an elaboration of my "Current Soviet Work in Sociology," *American Sociologist*, Vol. 1, No. 3 (May 1966). Some of the conclusions here differ from earlier ones, and I have added a good deal of data.

mobilizing and legitimizing aspects of the doctrine of Marxism-Leninism.

The continuing Soviet goal of a total bond between science and ideology is probably unique in the annals of modern scholarship. Within this framework, however, a partial shift in emphasis has taken place since the Stalin era. The shift has been most marked in the natural sciences; in the social sciences, by far the biggest change of late has been in the field of economics.

In comparison with economics, the "new sociology" lags a good deal in its "secularization"–that is, in the degree to which a field moves from sacred to profane ways of thinking. True, in the Soviet setting both of these social sciences share an intimate link with the state. Yet economics has a long record in Russia as a separate field of scholarship; sociology has none or almost none. Economics kept an identity during the Stalin era; sociology did not. In all societies, economics deals with the allocation of resources and hence efficiency; sociology deals with social relationships and hence social control. This makes modern economics inherently utilitarian and technical. Sociology, except for its most applied domains, is still highly normative and macroscopic in regard to societies as a whole. In the Soviet Union, most academic economists work with production officials; scholars in and around sociology tend to work with ideology officials.

These facts point not only to the limits but also to the significance of the new standing of Soviet sociology. As in the case of economics, this new standing did not emerge without a drawn-out debate involving scholars and officials. Whereas the debate on economics had to do with output–how to measure, plan, and raise it–the debate on sociology was more specific and also more esoteric. It revolved around the relation of the field to philosophy. In the Soviet setting this meant its relationship to the doctrine of Marxism-Leninism as well. Much of the published comment dealt with how "sociology" should be defined.

By 1966, this debate seemed to have been settled. Officials have apparently decided to link "sociology" to specialized social research, in the sense of methodically organized factual studies and of generalizations closely tied to such studies. At the same time, officials began to treat this kind of sociology as a field distinct from philosophy. The carrying out of this unstated decision was swift. It is no less unusual that we can trace the implementation step by step, by following a series of actions and statements.

A significant clue was provided in *Pravda*, almost in passing:

> Everyone knows of the rather melancholy experience connected with the development of concrete sociological research. This question was raised as far back as the middle 1950s. But at that time there were people who focused their attention on discussion of the legitimacy or illegitimacy of the term "sociology," and as a result the organization of concrete sociological research was delayed for a long time.[6]

Among the things that ended the debate was evidently a decision to leave open the matter of how "sociology" should be defined. This is evident in a 1964 speech by the man who was then the Communist Party's top ideology official, Leonid F. Il'ichev. He put forth an open-ended formula that a number of Soviet sociologists have since quoted:

> At present there exist among both Soviet and foreign Marxist scholars varying views on [the subject matter of] Marxist sociology. Some hold that Marxism as a whole makes up our sociology. Others link Marxist sociology

with the theory of scientific [building of] communism. Still others believe that between historical materialism and Marxist sociology there exists roughly the same contrast as between theoretical and applied areas of the same branch of human knowledge. One can find other shadings in outlook as well. In a word, as yet we have no common point of view.

So what about it? Should we forbid the representatives of this or that point of view the right to defend their opinion? Hardly. . . .

We must give up intolerant attitudes toward the opinions of each other, all the more since each of the points of view mentioned has in it a rational core which cannot be simply brushed aside.[7]

At the turn of 1966, no less than five events took place which embody the formal linking of sociology with specialized social research. To begin, late in 1965 a trial issue of the first Soviet journal in sociology was printed; its name is *Social Research (Sotsialnye issledovaniia)*. Second, the first national meeting of Soviet sociologists was held in Leningrad in February 1966. Some 600 people took part in it. Once again, the name of this meeting reflects the official decision about sociology: "Symposium on the Experience of Carrying out Concrete Sociological Research in the U.S.S.R." In the same month, the leadership of the Soviet Sociological Association passed from the veteran philosophers and ideologists who set it up in the mid-1950s to the younger sociologists most active in specialized social research. The Association's new president is Professor Gennadi V. Osipov, of Moscow. Osipov heads the social research unit of the top-level Philosophy Institute of the U.S.S.R. Academy of Sciences, and he writes widely.

As a fourth step, in February the Presidium of the U.S.S.R. Academy of Sciences adopted a special decision on "steps to improve the organization and coordination of concrete social research." It set up a new top-level body, the Learned Council on Problems of Concrete Social Research. According to a recent Soviet article, this Council has a mammoth task: "It has been charged with the coordination of [all] the concrete social research, including sociological research, that is being conducted in the country." It is interesting that the chairman of the Council is Academician Aleksei M. Rumiantsev. He now heads the Economics Department of the U.S.S.R. Academy of Sciences and is a member of the Party's Central Committee. During the past decade, Rumiantsev has become known as a leading ideologist of the post-Stalin generation. He was a long-time editor of the Soviet-sponsored international communist journal in Prague, *Problems of Peace and Socialism* (also called *World Marxist Review*), and then, more briefly, he was editor of *Pravda*.*

At the same time, the Academy's Presidium upgraded the main organizational base of Soviet sociology, in the Institute of Philosophy—heretofore the section on "labor and daily life"—into a Social Research Division. Units for concrete social research were set up in a number of the Academy's institutes, such as those on Ethnography, and State and Law. The Presidium concluded that the new Social Research Division of the Institute of Philosophy was to serve as the basis for a future separate Institute for Concrete Social and Social-Psychological Research.[8]

Lastly, a new Institute of the World Labor Movement came into being which set up its own Social Research Division. While the sociologists in the Institute of Philosophy focus in the main on social research within the

---

* In the Spring of 1967, the U.S.S.R. Academy of Sciences made Rumiantsev vice president for social science. Himself an economist, he took the place of a veteran philosopher and ideologist, Piotr N. Fedoseev.

Soviet Union, this second national center for the field seeks to stress and develop comparative studies, both of communist and noncommunist countries. Eduard A. Arab-Ogly heads the Social Research Division of the Institute. Among its members are some of the country's best-known sociologists, such as Professor Yuri A. Zamoshkin. This Institute combines an administrative base outside the U.S.S.R. Academy of Sciences (its funds come from the All-Union Central Council of Trade Unions) with a formal tie to the Academy's Economics Department under Academician Rumiantsev.

In view of the close Soviet link between the social sciences and an established ideology, it is no wonder that the new status of sociology received final sanction at a major Party event. I refer to the Twenty-third Congress of the Communist Party of the Soviet Union, held in March–April 1966. Here the new sociology was given a clear-cut status and definition that it never had before. At the outset of the Congress, the general report by the Party's leader cited sociology as a distinct social science. Leonid Brezhnev stressed its practical and applied mission, together with that of such other social sciences as economics:

> To elaborate important problems of economics, politics, philosophy, sociology, history, law, and other social sciences in close tie with the practice of communist construction is a most important task of Soviet scientists. . . . It is necessary to put an end to the idea, prevalent among a portion of our cadres, that they are called upon merely to explain and comment upon practice. The development of the social sciences and the implementation of their recommendations are of no less importance than the utilization of the achievements of the natural sciences in the sphere of material production and development of the people's spiritual life.

The head of the Soviet government echoed and made more specific Brezhnev's words about sociology. According to Aleksei Kosygin:

> Sociological research, based on a materialist understanding of history and generalizing the concrete facts of life of socialist society, is day by day playing a bigger role in the solution of practical questions of politics, production, and education.[9]

Soon after the Party Congress, three pleas for sociology made clear that the field was now getting added support. One kind of support came from a top Soviet mathematician and physicist, Academician Aleksandr D. Aleksandrov. Aleksandrov had long been rector of Leningrad University and now heads a research institute in Novosibirsk. Before the Congress, the leading younger sociologists had urged bigger and more autonomous training and research facilities for their work;[10] now the same plea came from a senior scholar. If such support had come earlier from some of the leading philosophers, the background of Academician Aleksandrov and the practical theme of his plea testify to the new status of Soviet sociology. The same can be said of an article in *Komsomol'-skaia Pravda* which reflects a special interest of top youth officials in the uses of sociology. And another long article urges Soviet writers to look to sociology as a "kindred" science.[11]

In these ways, specialized quantitative research and advanced methods for conducting such research have won a legitimation they did not possess before. The same is true of sociology as a discipline more or less distinct from philosophy and Marxism-Leninism as such. The next step is to weigh just what the new status of Soviet sociology might mean in regard to the interaction between science and ideology.

The "Future" Orientation

The "future" orientation stands much closer to the earlier
Soviet deductive social philosophy. It flows from a lasting
feature of Soviet ideology: an official long-range blue-
print for moving into a more or less perfect future. Here,
the over-all analysis of society has as its frame of reference
not the present but an officially defined future. Within
the field of sociology, many books now mirror this future
orientation of established ideology. These books differ a
great deal from leading "concrete" studies based on specialized social research, and in the mid-1960s they were
quite a bit more numerous than the latter.

Aside from the emphasis on a long-range official blue-
print, three traits set these works apart from the "concrete" studies. They draw their data from a melange of
casually cited sources, which range from census and
output data to anecdotes from the mass media. Quotations
from official documents and doctrinal classics stand out
as the main evidence. From time to time, small pieces of
quantitative survey research (interviews and questionnaires) are added. One study, for example, gave such
data to support the point that skilled workers take part
in local administration more than do other social groups.
The three rows of figures in Table 1 seem to support the
point. But the evidence is most limited in amount and
scope and, furthermore, the reader learns nothing at all
about how the data were put together.

Second, the topics of these studies come much closer
to the central institutions of the society than do the topics
of "concrete" research. This is true of all four studies I
shall mention. Thus, Anatoli A. Zvorykin, until recently a
member of the Philosophy Institute and a veteran student
of science and technology, takes up the nature and social

Table 1. *Collective farmers who take part in local administration (in per cent)*[a]

Name of collective farm[b]	Mechanics	Animal husbandry	Construction and maintenance	Farm field work
"XXI Congress of CPSU"	23	12	8	5
(N = 1359)	(177)	(210)	(60)	(912)
"Lenin"	14	9	9	4
(N = 1564)	(257)	(167)	(32)	(1108)
"Young Communist"	29	10	9	4
(N = 273)	(31)	(72)	(11)	(159)

[a] The percentages are based on totals given below them.

[b] Located in Tomsk Province, in Siberia. "N" refers to the number of members in each collective farm.

Source: Adapted from table in Vladimir I. Razin, *Stanovlenie kommunisticheskogo samoupravleniia* [The Development of Communist Self-Government] (Moscow: Moscow University Press, 1965), p. 174.

effects of technology. Another book reports an attempt to study an entire community, a Moldavian (ex-Rumanian) village named Kopanka. Although this study includes only the least analytical type of quantitative research, two survey specialists—Osipov and Shubkin—were among its project directors. A group of sociologists working with Vladimir I. Razin, of Moscow University, wrote a book on the trends toward popular participation in government. Still another book with a future orientation was put out under the auspices of the Party's own Academy of Social Sciences; its topic reaches the very heart of classic Marxist sociology—the make-up and relations of classes.[12]

These studies have one more trait in common. All of them talk of today's society as no more than a passing way station en route to an ideal though distant future. What makes progress certain are above all else the boun- ties of science and technology. This remains a major theme in post-Stalin doctrines about the future. In particular, those sociologists with a future orientation echo it time and again. Any fitting elements in the here and now not only are singled out but are portrayed with euphoria. Despite their broad topics, then, these books tell very little about any part of present-day reality. They recall the traditional Soviet ideology much more than they do a scientific study of society.

The "Concrete" Orientation

If the future orientation calls to mind the deductive and hence dogmatic element in Soviet ideology, the second, concrete orientation cannot be said to run counter to the established ideology. Rather, it fits a quite distinct ele- ment within the same ideology. This element puts great value on "practice," and on a full knowledge of what is going on in the real world. Under Stalin, this element of

Soviet ideology fell on bad times, and so did the systematic study of society for which it seems to call. In recent years, and in part only since the fall of Khrushchev,
this "realistic" element in ideology has made something
of a comeback. This brings us to a key point. The concrete
orientation within sociology, which in various ways stands
closer to science than does the future one, owes its gains
not to a conflict with ideology but to harmony with its
more "realistic" element.

There has been a steady rise in the status and amount
of "concrete" social research. That was already apparent
during the decade prior to the 1966 shift described above.
Technically, this specialized type of research means a
stress on quantitative methods: assembling firsthand data
by means of systematic "surveys" (interviews or questionnaires), and analyzing such data with rigorous statistical
methods. More broadly, the concrete orientation seeks to
use methodical quantitative research to bear out any
generalizations about social reality.

A strong Soviet emphasis on such research does not
call for as much change in outlook as might be expected.
At all times, as Ralf Dahrendorf has noted, "the party
organization and its varied affiliations serve as a gigantic
institute of opinion research. . . ."[18] In this context, and
also in line with the less deductive element in ideology,
the post-Stalin state favors both mass opinion-polling and
scholarly survey research as added channels of information and communication. Many scholars, too, welcome
"concrete sociological research" as an advanced technique.[14]

In nonacademic polling, the Public Opinion Institute
of the main daily paper for youth, *Komsomol'skaia Pravda,*
still holds a leading place. Thus their poll "replicated" a
Gallup poll on how youth views life. But by now other
polls appear as well. In late 1964, for instance, the central

press told of at least three polls: attitudes to village life, to army life, and to theater-going.[15]

In scholarly research, the scope has widened from attitudes toward work to include occupational choice as well as the uses of time outside of work.[16] The major Soviet output of "concrete" survey research consists of three major studies. Started first was a study in Gorky, a large industrial city on the Volga. Of the two other studies, one was conducted in Leningrad, the other in the "Chicago of Siberia," Novosibirsk. So far, the Gorky and Novosibirsk studies have published a full account of their own work in a book. The Leningrad study reportedly plans to do the same soon. Meanwhile, it has put out an interim report in book form and also some articles.*

The Gorky study, headed by Osipov, was a cooperative project of sociologists at the Philosophy Institute in Moscow and of various social scientists in Gorky. The study drew its data from five factories in or near Gorky. Many workers (as well as engineers and technicians) were interviewed, and the resultant book contains numerous statistical tables and other facts. But often the tables do not state the sample on which they are based. In general, one gets no clear picture of how many people were studied, or just how they were selected.

The Gorky study stresses the impact of technological change on factory workers' lives and work. It gives more data than do the other studies on material factors—es-

---

* The 392-page final report of the Leningrad study, *Chelovek i ego rabota* [Man and His Work] (Moscow: "Mysl," 1967), came out as this book went to press. The main authors are Andrei G. Zdravomyslov and Vladimir A. Yadov. The study shows more progress, analytical as well as technical, than any other Soviet work in sociology and, in general, it stands out as a major book.

A large appendix contains summary tables of quantitative data and technical statements on the collection and statistical analysis of the data.

pecially types of technology and levels of skill—and less data about either objective or subjective social factors.[17]

The Leningrad study was led by the director of the Social Research Laboratory of Leningrad University, Vladimir A. Yadov, and by Andrei G. Zdravomyslov of the Leningrad branch of the U.S.S.R. Academy of Sciences. Much of the study draws on a sizable questionnaire, filled out by a stratified random sample of 2,665 workers under the age of thirty. Findings cover such topics as their technical creativity (the "rationalization" proposals made to raise output), attitudes toward manual occupations, how levels of skill and education affect attitudes to work, reasons for changing jobs, and value orientations toward work and leisure.[18]

The Novosibirsk study is a pioneer work in Soviet analysis of social mobility. The study deals with careers young people want and those they in fact start on. It was headed by Vladimir N. Shubkin, chief of the social research section of a Laboratory of Mathematical Economics at Novosibirsk University. Like the one in Leningrad, the Novosibirsk study relies in large measure on individual questionnaires. In the spring of 1963, for instance, one of a series of half a dozen such questionnaires was filled out by a simple random sample; 66 per cent of the high school seniors of Novosibirsk Province made up the sample. (The questionnaire went to a universe of 4,427; 2,940 responded.) A follow-up survey asked for information from school principals on initial career steps of the respondents.[19]

In terms of specific techniques, the Novosibirsk study relied on open-ended questions; the Leningrad one, on a more complex set of pretested multiple-choice questions. The same contrast appears in the analysis of data. The measure of association between variables used in the Novosibirsk study was a simple correlation. The Leningrad study also used partial correlations and a range of significance tests, including chi-square. In all, on the technical level the Leningrad study seems to rank higher than the Novosibirsk one, with the Gorky study somewhere in between. The latter used both open-ended questions and closed, "fixed alternative" questions. Its mathematical measures of associations between variables relied on multiple correlation techniques. But the Gorky study does not give a clear picture of its sample. Hence one cannot compare the sampling techniques used.

The methods just sketched are open to challenge on specific technical points. Beyond that, they show how new the advanced methods of quantitative research are to Soviet sociologists. At the same time, the ever greater use of such methods by those who follow the concrete orientation indicates a strong bond to science and its ways.

In weighing the compatibility of a universal science with a monist ideology, however, the methods used are not the only point of interest in the new Soviet quantitative research. Even more relevant is the content. The social scientists of any country face the question of how well, how fully and honestly, they show the social reality. In the Soviet case, long years of fulsome idealization and outright lies make the question all the more pertinent. Can it be that Soviet scholars are now making public social data that could go counter to an official, often still heavily idealized, picture of life?

I shall return to this question when I speak of Soviet theoretical interest in the Marxian problem of work and alienation. Meanwhile let it be said that, as the three big surveys show, some Soviet sociologists do get much closer to gray and even black shadings of reality now than at any time since the 1920s. True, this is not always so; the cultural style remains ideological much of the time. But in the last few years a shift has been apparent even here.

This may be illustrated by a closer look at the Leningrad and Novosibirsk studies.*

In the Leningrad study, the authors keep coming back to a central question: What factors help to make work a rich, humanizing experience, and what factors stand in the way? A central essay by the study's co-directors, Zdravomyslov and Yadov, weighs the effect that basic value orientations have on the work attitudes of young workers in Leningrad. At the outset, they look at three aspects of work. They single out performance in one's job as an objective aspect, while work satisfaction and motivation constitute two subjective aspects.

The authors evaluate work performance by means of an index of six facts about how a man works. Three deal with how effective (efficient and reliable) a man is on his job, and three with how much individual initiative he shows (such as making proposals for technical "rationalizations"). For job satisfaction, they offer an index based on answers to three questions: Do you like your present work? Would you like to move to another job? If for some reason you do not work for a while, would you want to go back to your present work? Lastly, on work motivation, they set apart those individuals who put service to society first, with or without an added motivation, and those who, one way or the other, put first their pay and making money.

The Leningrad study uses leisure, too, to study the value orientations. Each worker was asked what he now likes to do most. He was also asked what he would like to do best if he had more leisure time. If a respondent chose the same pursuit both times, this was defined as

---

*I omit comment on the Gorky study. For the issue at hand, its technology-oriented content has less relevance than do the other studies. On theoretical trends, however, it is the Gorky study that tells us the most of all.

his primary orientation. Among those with this orientation, by far the largest group chose the family. The next biggest group chose education; much smaller groups chose civic work, their job, and making money. (See Table 2.)

Table 2. Interaction of value orientations among young Soviet workers

(Percentage of Secondary orientations for each Primary orientation)

Primary orientation		Secondary orientations				
	Number	*Family*	*Edu-cation*	*Civic work*	*Job*	*Pay*
Family	(1,100)	—	15	14	10	10
Education	(627)	23	—	20	15	5
Civic work	(329)	42	42	—	21	7
Job	(207)	55	48	37	—	12
Pay	(161)	66	21	15	16	—
No orientation	(970)	—	—	—	—	—
Distribution in whole sample	(2,667)	38	23	12	10	6

Source: Adapted from table in Andrei G. Zdravomyslov and Vladimir A. Yadov, "Attitudes to Work and Personal Value Orientations" (in Russian), in G. V. Osipov, ed., *Sotsiologiia v SSSR* [Sociology in the USSR], Vol. II (Moscow: "Nauka," 1965), 205.

Family life is not only the most favored primary orientation but it is also the main second choice among those with other primary orientations.

As a final step, the study relates the primary value orientations of young Leningrad workers to the various work attitudes. Table 3 summarizes the more detailed data. The bottom row indicates how the authors ranked the value orientations in terms of their positive impact. As could be expected, those oriented toward their job had the most favorable specific attitudes as well. A prime interest in family life, however, led to no more than a middling set of work attitudes. So did education, which possibly meant, according to the authors, a prime interest in a different, better job.

Table 3.   The effect of value orientations on attitudes to work

(How the attitudes of each Primary orientation compare to the sample as a whole)

Aspect of work attitudes	Primary orientation					
	Family	Edu- cation	Civic work	Job	Pay	No orien- tation
Work orientation	+	0	+	+	0	—
Work satisfaction	+	0	0	+	—	+
Work motivation (service)	—	+	+	+	—	—
Positive effect of primary orientation on attitudes: rank	4	3	2	1	5	6

Note: + The effect is greater than on the sample.
  — The effect is less than on the sample.
  0 The effect is roughly the same as on the sample.

Source: Same as Table 2, adapted from table on p. 206 (based on average of indexes).

Zdravomyslov and Yadov end their article with a revealing passage. The context is the discussion of an upbringing which would assure "communist attitudes toward work." This closing passage tells us a good deal about how leading Soviet sociologists now speak of the practical problems raised by their research:

The first problem flows from the fact that quite a large part of the young workers turn out to have no clear-cut orientation. Systematic work is needed especially with this group. [This work] could begin by overcoming a feeling of complacency, by [implanting] on the contrary a discontent with one's activities through a harsh evaluation of the results of such work. At the same time further work is needed to clarify the actual orientations of just this part of the youth.

A second problem consists of making clear the social significance of work, and the inculcation on this basis of a feeling of duty, in that part of the youth which orients itself toward making money. Here it is still more essential to find out the interests outside production and the forms of using one's free time, and also to make clear the genuine values of human culture and to cultivate respect toward them.[20]

The findings of the Novosibirsk study were highlighted in a 1965 essay by Shubkin which appeared in the leading Soviet journal of philosophy and sociology. *Voprosy filosofii* noted that "this article is published by way of discussion," and the article got quite a bit of attention in the Soviet Union. Entitled "Youth Enters Life," it speaks with candor about a wide range of problems that Soviet young people face when they come out of school. Shubkin contends that a large double gap exists. For one thing, many young people end up working in fields other than those in which they received vocational training. But even more important, says the author, is the gap between plans to go on to higher education and the necessity to start work right away:

> For instance, 80 per cent of the graduates [of secondary schools] intended to continue their studies. These personal plans were substantially changed by objective conditions. Not 80 per cent but 44 per cent of the graduates went straight from secondary school to higher studies. . . . A different picture emerges when we look at the personal plans of those graduates who intended to go to work in the national economy. Only 8 per cent intended to go to

work immediately. Actually, not 8 per cent but 32 per cent
went straight from school to work. But, as was noted above,
the majority of graduates do not go to work in the specialty
for which they were trained in school.[21]

To Shubkin, the main problem perhaps is a social one.
To show this, the Novosibirsk study presents data on a
cross-section of urban and rural strata. Thus it goes beyond
the usual Soviet categories of workers and, at times,
professionals and subprofessionals in industry. With a
variety of quantitative data, Shubkin gives a wholly un-
ambiguous interpretation of these data, which points to
the differences among social strata that are often known
but not given their due. This is how Shubkin puts it:

> It is important to keep in mind that opportunities for en-
> tering higher schools depend substantially not only on the
> demographic situation in the country [variations between
> regions and the "baby boom" after World War II] but
> also on a number of social factors which, in our view, it
> is impermissible to ignore. At present, young people who
> live in large cities have a substantially greater chance of
> entering a higher school than rural youngsters, since as a
> rule the level of preparation is higher in the city schools.
> Analysis also indicates that the level of education of the
> parents has a strong influence on the nature of the chil-
> dren's interests and aspirations. Advancement, vocational
> preferences and, consequently, educational opportunities
> are greatly influenced by material and housing conditions,
> and so forth. Because of all this, the social structure of the
> body of children who enter first grade is noticeably differ-
> ent from the social composition of the student body of
> higher educational institutions.[22]

Shubkin supports these statements with other specific
data on how social origin affects educational opportunities. As can be seen in Table 4, aspirations and actual

Table 4. Social origin and early careers of Soviet high school graduates (in per cent)

Group to which parents belong	Proportion of graduates wishing to:			Proportion of graduates who did:		
	Work	Combine work with study	Study	Work	Combine work with study	Study
Urban professionals	2	5	93	15	3	82
Rural professionals	11	13	76	42	—	58
Sub-professionals and workers in industry and construction	11	6	83	36	3	61
Sub-professionals and workers in transport and communications	—	18	82	55	—	45
Sub-professionals and workers in agriculture	10	14	76	90	—	10
Sub-professionals and workers in service work	9	15	76	38	3	59
Other	12	38	50	63	12	25
Percentage of the whole sample N = 2940	7	10	83	37	2	61

Source: Adapted from tables in Vladimir N. Shubkin, ". . . Problems of Job Placement and Occupational Choice" (in Russian), *Voprosy filosofii* XIX, 5 (May 1965), translated in CDSP, XVII, 30 (August 13, 1965), 6.

opportunities come closest among children of urban professionals and are furthest apart for children of manual workers, especially in agriculture. Shubkin makes this clear:

> While we are categorically opposed to an overly broad interpretation of these selective data, we nonetheless have a basis for asserting that the paths in life [career opportunities] of young people in various social groups at present differ substantially. Of 100 graduates from families of agricultural workers [state farm workers and collective farmers], only 10 continued their studies after completing secondary school, while 90 went to work; of 100 graduates from families of the urban intelligentsia, 82 continued their studies and only 15 went to work [the other 3 combined work with study].[23]

Shubkin's other major point concerns occupational preference. He documents contrasts between urban and rural youth and between boys and girls. He argues throughout that "particular attention should be paid to the low attractiveness of mass occupations in agriculture and service trades." On the steps needed to change this, Shubkin adds parenthetically: "We are speaking, of course, not of administrative but of socio-economic and other measures."

The Novosibirsk study bases its analysis on a poll of occupational preference, which is unique for the Soviet Union. The ratings assigned by school graduates to seventy typical occupations appear in the 1964 volume reporting that study. In his 1965 essay, Shubkin sums up the findings on two major occupational groups: professional academic work and all subprofessional and manual work. Among professional careers, the natural sciences rank on the whole much higher than the humanities and social sciences. In the other group, the sharpest contrast lies between the low-ranking service work and the rest (see Table 5).

Table 5. Occupational preference[a] of Soviet high school graduates

Professional work in academic fields[b]

	Rank	Points
Physics	1	7.69
Mathematics	2	7.50
Medical research	3	7.32
Chemistry	4	7.23
Geology	5	6.84
Mathematical economics	6	6.33
History	7	6.17
Philosophy	8	6.06
Philology	9	5.75
[General] economics	10	5.52
Biology	11	4.66

Sub-professional and manual work

	Rank	Points
Transport and communications[c]	1	5.28
Education, culture, and public health	2	4.82
Industry	3	4.26
Construction	4	4.07
Agriculture	5	3.75
Services (e.g., sales and clerical)	6	2.63

[a] Evaluation on point scale from 1 to 10 (N=280).

[b] Outside science, three occupations fall between ranks 3 and 4: doctors, writers, and artists. The mean rating of fourteen engineering occupations puts engineers, as a group, between ranks 6 and 7.

[c] The occupations of pilot and radio mechanic received the second-highest rating in the whole survey, just below physics.

Source: Same as Table 4.

In closing, Shubkin relates his topic to the Soviet system, on the one hand, and to the mission of the "new sociology," on the other:

> The investigation we carried out shows that Soviet boys and girls starting out in life and work are faced with many complicated problems. At the same time, experience shows that in a socialist society every possibility exists to solve the contradictions and overcome the difficulties that arise. But to act we must know. Therefore one of the most urgent and effective forms of assistance to young people by social scientists is the broad development of sociological research and the elaboration of concrete proposals for the directing bodies of our society.[24]

What, then, are the Soviet ways of doing quantitative social research? In the new studies of Leningrad, Novosibirsk, and Gorky, some advanced methods and techniques are used to illustrate the complex issues of Marxian theory that I discuss below. Aside from these three studies, and to some extent within them, recent research has still been technically limited on several counts.

Soviet sociologists tend to pay much more heed to the gathering and measuring of quantitative data than to its analysis and interpretation. Nearly all of the current work belongs to the descriptive type of survey and does not attempt explanatory, "causal" research. As yet no one has tried anything like the structural and "relational" analysis, for instance, of which Coleman speaks.[25] Most of the data focus on attitudes rather than on objective traits of individuals or groups. This is why Parsons could conclude after a visit with Soviet sociologists that "a considerable part of the group's empirical research would be classified by American sociologists as social psychology."[26] Finally, on the selected social problems that Soviet sociologists study, the policy recommendations they publish neither say nor imply that all is well at present. Yet these recommendations often turn out to be quite unsystematic and lacking in novelty.

The partly high technical level of the three major studies suggests that some Soviet specialists welcome systematic quantitative research. These studies leave little doubt as to the active commitment to quantitative research and its advanced forms on the part of sociologists with a concrete orientation.

The matter of content is more complex. I note later some of the problems that stand in the way of knowing and weighing current Soviet work. Any view of social research depends in large measure on its theoretical base, a no less complex matter to which I shall later turn. Within these limits one thing appears to be clear about the data and interpretations cited here. The three new Soviet studies investigate some parts of social reality quite systematically and fully. And they do so very much in line with one of the elements of established ideology, the "realistic" one. Within this context, at least, a science-minded orientation in the new sociology shows more harmony than conflict with current ideology.

Criticisms of Western Sociology

On no other subject have so many leading Soviet scholars in the field written books as on Western or "bourgeois" sociology. The more solid of these books contain documentation and show some real knowledge.[27] Most of them treat specific questions.

As far back as 1959, Vadim S. Semenov, of the Philosophy Institute, published a volume on "problems of classes and class struggle in contemporary bourgeois sociology." In 1962, the only book by a Soviet sociologist, Dzherman M. Gvishiani, on problems of management

appeared—*The Sociology of [American] Business* (in Russian). Two years later, sociologists at Leningrad University put out two specialized studies on Western social
theory. Igor S. Kon wrote on positivism and neo-positivism,
and Andrei G. Zdravomyslov on the concept of interests.
In the one recent Soviet book on religion and its general
social setting (1965), Yuri A. Levada, a theorist at the
Philosophy Institute, deals almost wholly with Western
theory and research. In 1966, two members of the new
Institute on the World Labor Movement published books
on American sociology: Yuri A. Zamoshkin, *The Crisis of
Bourgeois Individualism and Personality*, and Nikolai V.
Novikov, *A Critique of the Bourgeois "Science of Social
Behavior"* (both in Russian).

Two major works, however, treat the subject of Western sociology as a whole. One is by Osipov, the other by
Galina M. Andreeva, of Moscow University.[28] Osipov's
book, which came out in 1964, deals with the main concepts of Western sociology. It discusses three sets of concepts: society, social classes and groups, social change and
social progress. Andreeva's work, on the other hand, centers on the methods of empirical research. Her book,
published in 1965, has chapters on the main fields of
empirical research, on the procedure and techniques of
research, and on the main themes of empirical methodology. A concluding chapter contains a critique of Robert
Merton's case for theories of the middle range and of
Talcott Parsons' theory of action.

Andreeva's book seems more systematic. It gives the
reader more relevant facts, and it has fewer bibliographic
errors. Andreeva does not rely on quotations from Marx
and Lenin (and from third-rate American textbooks) as
often or as extensively as does Osipov.

Apart from these contrasts, though, the two books
have much in common with each other and with other

Soviet works on Western sociology. Most revealing is the over-all mode of writing about the ways and means of sociology. A Manichean dichotomy overshadows all else —one side embodies darkness and failure, the other is all light and success. True, from time to time a line is drawn between more "enlightened" scholars in the West and those beyond redemption. Likewise, note is taken of tech nical advances, and of some "valid" factual findings, which Marxists can and should use. The Manichean phase may have passed its peak. This is not true, though, of anything printed about theory, general methodology, and over-all ideas about reality. In these works, the tone continues to be one of fierce and total combat, of true believers prepared to battle the infidels to the bitter end.

In Mannheim's terms, this mode of writing reflects both a "total" conception of ideology and a "special" form of that conception.[29] In the case of any opponent, not just some ideas but all of them must be declared, perforce, to be "ideological," in the negative sense of false consciousness. It is not every group but only an opponent who is bound by ideas that mirror his group's place in society. In her book, Andreeva shows how Soviet sociologists put these "total" and "special" views of ideology today:

> The values recognized in a society are, of course, always values of definite social forces. Bourgeois sociology does not see a possible objective criterion for weighing the values themselves. But that is just the problem. Such a criterion does exist. It reveals itself in social science if it bases itself on certain philosophical principles—on the principles of that world view which is the world view of the leading progressive force of history in the contemporary era, the working class.
>
> Any other philosophy cannot, indeed, offer objective criteria of values. The approach to values which marks, for instance, the philosophy of neo-Kantianism produced

nothing except a new variant of subjectivism. In rejecting the "speculations" of philosophy, empirical sociology rejects along with them the value approach. It is right in that part where it admits the fruitlessness of the approach of idealistic philosophy. It is not right when it does not see the possibility of a fundamentally different approach to values within philosophy itself. But on this question [empirical sociology] cannot be right, for its class and theoretical position rules out such a possibility.[30]

What accounts for the strong emphasis on criticizing Western sociology and for this mode of doing it? One likely reason is the strain attendant upon the swift acculturation and legitimation of science. That strain must be great. Not only does it involve a general setting, within which many Soviet scholars carry out official "combat" missions against the West, but it also expresses a strong national urge of Soviet sociologists to set themselves off clearly from other kinds of sociology, and to push claims for their own. For Soviet scholars in the field and for their readers, this "combat" has a latent cognitive function. In the course of doing battle, they learn and tell a good deal about what scientists are doing elsewhere. All the same, both the official mission and a national urge lead to strife, especially with American sociology.

Equally salient is an outlook which Soviet sociologists find all too rare in the West. To judge by their writings, all members of the profession share a strong, unambivalent rationalism. Theirs remains a highly optimistic and activist outlook. The outlook is collectivist and statist as well. Soviet scholars tend to believe in highly activist ways of solving social problems. They see this activism partly as a matter of ubiquitous leadership by a total state and partly as a result of their own "committed" scientific work.

The lack of any such outlook among most Western sociologists makes them seem alien and their work dubious in Soviet eyes.[31]

Lastly, in the Soviet Union sociologists (and presumably their publishers) share with party officials an ideological view of how societies should be either organized or analyzed. On matters of doctrine, moreover, Soviet ways of arguing and writing have changed much less than they have on many other things.

Thus the often savage Soviet criticism of Western sociology comes from men and women who believe deeply in what Mannheim calls the "total" and "special" views of ideology. In cultural style, too, Soviet sociology remains much closer to ideology. Merton speaks of such battles as "scientific controversies." He reminds us that they are common to all sociologists, no matter whether or not their ideologies differ.[32]

Scholarly disputes between Soviet sociologists and their colleagues abroad continue to suffer from major handicaps. Before such disputes can become more or less systematic, sociologists abroad need to know much more about Soviet work in the field and about the reality they portray. Several obstacles now stand in the way. For one thing, Soviet scholars have not even begun to make available to their colleagues the kind of detailed data on findings and research tools that many sociologists elsewhere exchange with each other all the time. Nor can foreign scholars carry out firsthand social research in the Soviet Union, as they do in many other societies today. Also, foreign sociologists have not yet discussed the recent Soviet survey studies fully in any scholarly international forum. This would offer a context within which to weigh their substantive and methodological competence. Beyond that, it may take some time before we can expect any major Soviet discussions of their own studies, for instance, along the lines of the well-known *Continuities in Social Research*. Meanwhile, Soviet "first steps" are the subject of only brief comments at home and abroad. Together with the ideological style of Soviet criticism of Western work, these handicaps make it all the harder to draw the line between ideological combat and scientific controversy.

Marx, Work, and Alienation

Science involves more than data, and more than more-or-less systematic methods of assembling and analyzing them. The search for basic knowledge begins and ends with theory. Whether it takes the form of limited generalizations or universal laws, it is the quality of its theorizing that gives a science its meaning and standing.

In Western eyes this raises some important issues about Soviet sociology. Its theory seems to stay wedded to a "closed" system of universal laws—the Stalinist form of Marxism. To many non-Soviet scholars, Stalin stands for a mummified version of a century-old view. Nor was Marxism-Leninism "closed" only in the logical sense of the word. Under Stalin, it came to mean a more or less total tie between science and ideology. When one adds that an official ideology still rules over science, what kind of theorizing can Soviet sociologists do? These are big issues indeed.

The answer rests not so much in this or that form of Marxism as in a changing social setting. To this day, the Stalinist heritage lies heavy on all of Soviet life. Yet, as the Soviet system marks its fiftieth birthday, we can see it going through at least two important changes. Greater economic affluence makes the individual Soviet citizen matter a good deal more, as both producer and consumer, than he did during Stalin's age of "primitive accumulation." At the same time, a significant shift is occurring in politics, from absolute totalitarianism to more enlightened rule. This shift creates the need for a smoother popular compliance with official policies than was necessary in the Stalin era. In turn, this need sets off pressures to replace a ubiquitous secret police, a mass-persuasion system, and the like, with new mechanisms of information, communication, and social control.

In their current work, Soviet sociologists put much emphasis on one of these great changes but not on the other: they deal largely not with the social system as a whole but with the individual and his interaction with a more and more complex economy. It is revealing of the social context of post-Stalin sociology that in the realm of theory these potentially interesting "first steps" stop short of the realm of politics. Just how far this goes may be seen from a recent survey of what sociologists are doing in the communist countries of Eastern Europe. The survey was made for UNESCO by a well-established Polish sociologist, and it covers work in political sociology.[33] The author, Jerzy J. Wiatr, opens his survey with a proposition not echoed in the writing of Soviet sociologists. According to Wiatr, politics lies at the heart of what Marxist sociologists should and do work on:

> The study of politics becomes, therefore, the central point in the analysis of social relations; any scientific analysis of society is inconceivable without it. . . . As a result, the Marxist interpretation of social phenomena, the Marxist sociology becomes above all the sociology of political relations. . . . It is because of this political orientation of Marxism that in the Marxist social sciences politics is the supreme and essential aspect of social life.[34]

The new Soviet sociology puts no such stress on politics. On the level of research, the same gulf appears in regard to central Marxist concern with classes and class structure. In Eastern Europe, Wiatr writes:

> The Marxist theory of classes is given a broader interpretation by including such conceptions as the theory of social stratification and the theory of pressure groups. These two [theories], and the developments they are concerned with, make the Marxist political sociology face some phenomena which either came into being after the birth of Marxist theory or had not been studied until recently. . . . Discussions on the subject of differences in stratification and their evolution constitute one of the more lively and interesting trends in today's political sociology in the socialist countries.[35]

No Soviet sociologist today would write as Wiatr does. Wiatr not only comments that, "according to an old Marxist forecast, the socialist revolution will lead in the long run to the withering away of the bureaucratic apparatus of power which will be replaced by social self-government in various forms"; he also adds, "as we all know, that forecast has not come true; the problems of large bureaucratic structures, which preoccupy modern political sociology everywhere, exist in the socialist countries too."[36]

Another well-established Polish scholar, the social theorist Adam Schaff, remarks along the same lines:

> The anti-Communist literature writes . . . about a "new class," about "the bourgeoisie," etc. From a theoretical point of view, this is of course nonsensical: the elite that emerges in socialist countries is not a class *ex definitione*, and the use of the term "bourgeoisie" can be understood only as a cheap propaganda trick on the part of our enemies. But this should not console us: it is much easier to

reject the exploitation of a problem by our enemies than to reject the problem itself. And the problem consists in the fact that a power elite has been forming in socialist countries which in the natural course of events enjoys a privileged social position. This state of affairs is completely natural and socially justified. There is no reason for an embarrassed silence. However, while this state of affairs is natural and at a given stage of development even inevitable, it is at the same time pregnant with danger, and if for no other reason, therefore, it should be carefully examined and reflected upon. The danger is that an extension of privilege beyond certain limits and the failure to take counteractions against such tendencies lead from the point of view of socialism to negative consequences.[37]

A younger Polish sociologist, Aleksandra Jasinska, also writes of Marxist theory of the state in ways that no Soviet sociologists do:

> By their essence—didactic aims—propagandistic popularizations and textbooks induce the closed character of their systematization of Marxian theory. They leave no room for concepts and formulations requiring further explanation and greater precision, [for] adequately elaborated problems and propositions requiring verification.[38]

In contrast to the concept of "political sociology" to which Wiatr, Schaff, and Jasinska adhere, a statement by Andreeva takes a different view. At the Fifth World Congress of Sociology in 1962, Andreeva listed the studies she and other Soviet sociologists were conducting on intellectuals, professionals, and leadership groups. As yet, none of this research has come out in book form, and very little of it in any other form. At the same time, Andreeva gave a number of arguments explaining why Soviet sociologists rule out elite research. The arguments go from a rejection of the very concept of "elite" to a denial that in the Soviet Union political leaders can validly be termed a distinct leadership group or elite.[39]

Hence, when Soviet sociologists speak at all of politics and the Marxist theory of it, they limit themselves to current official themes.* This is not quite so when it comes to the realm of work. In this area, in a partial way, official themes and language now mix with basic questions of Marxist theory. Specifically, some evidence exists that leading "concrete" sociologists may be trying to fit the classic theme of alienation to Soviet society and its current stage of development. Along with modified Stalinist dogmas about the necessity for the system to remake man and society "from above," we may have here a novel and quite systematic linking of micro-research with classic macro-theory.

The fullest evidence of such a trend comes from the recent report on the large Gorky study, *The Working Class and Technical Progress* (in Russian).[40] One thing stands out, to begin with. The factor most emphasized in this study is technology and its changing impact on industrial labor. Many other factors are mentioned, and the data are not limited to technology and related technical skills. Yet it does not seem to be by chance alone that all three illustrations in the book—two photographs and the jacket design—show machines, not people.

* Shortly before this book went to press, a long review essay in *Voprosy filosofii* (XXI, 4, 1967) pointed to a new and broader view of political sociology. Written by V. V. Varchuk and V. I. Razin, the essay was translated in CDSP, XIX, 23 (June 28, 1967) under the title "Research in the Sphere of the Political Organization of Socialist Society." It found most recent Soviet writings on the subject weak, and came close to some of the Polish views cited here. The authors single out one Soviet work as looking at mass participation in public life both broadly and empirically: Yuri Ye. Volkov, *Tak rozhdaetsia kommunisticheskoe samoupravlenie* [How Communist Self Government is Born] (Moscow: "Mysl," 1965).

From the outset, then, we find the same stress on the life-transforming bounties of science and technology that marks so much of Soviet doctrine and research today. No less dominant—and no less optimistic—is another theme. Whereas capitalism can use such bounties no more fully now than it could in Marx's time, the Soviet social system is in an ideal position to use them to build a good society. On the subject of work, the Gorky study also fits the prevailing Soviet view. Under capitalism, automation and the like lead to pessimism and social disorganization; in the Soviet setting the new technology can and will assure a truly humane type of labor, creative and deeply satisfying.

Yet the book on the Gorky study does not stop with these common beliefs. Here and there, it goes further in adumbrating a new Soviet theory of work and alienation.[41] This lends special interest to what reads like a restatement of traditional views. The less standard ideas in *The Working Class and Technical Progress* can be set forth as a series of theoretical propositions:

1. *The division of labor leads to an important economic form of alienation.*

The problem of alienation is now seen to constitute a distinct and central problem in Marx's social theory. The writers state: "The problem of alienation was scientifically solved in the economic works of Karl Marx of the late 1850s and 1860s. The category of alienation holds one of the most important places in the Marxian conception of society. It is closely linked with such concepts as the division of labor, classes, social institutions, and so on. But at the same time it has a distinct content of its own and does not become subsumed by any of these concepts."[42] Since Marx sought to emphasize the economic bases of capitalist society, "it was natural that an important place

in his works is assigned to economic forms of alienation—
the division of labor, forms of property."[43]

This point brings us to the link between alienation and
the realm of work: "An inevitable consequence of an
antagonistic [capitalist] division of labor is alienation from
work, the alienation of man from himself in the producing
of work."[44] Marx spoke of three forms of such alienation
from work: the worker's view of his own work, his view
of the product of his work, and how he sees his place in
society. Most immediately, then, alienation from work
flows from the ubiquitous social division of labor.

2. *Technological and personality factors can cause
alienation from work in noncapitalist modern societies.*

In the Soviet Union, socialism eliminated the objective
cause of alienation from work, a capitalist or "antagonistic" division of labor based on private ownership of
the means of production. Yet the phenomenon of alienation has other aspects. The book on the Gorky study calls
them the "technical-economic and subjective aspects" of
alienation. A revealing passage makes clear that such aspects can be found in a modern society without capitalism
—including the Soviet Union:

> Not so long ago, these aspects of alienation from work
> were almost not studied. They were considered merely
> survivals [of capitalism], to be liquidated with the help
> of [communist] upbringing. No attention at all was given
> to the fact that [something else takes place] in socialist
> society in the course of the development and interaction
> of objective factors (the level of technological develop-
> ment, the immediate conditions of work, the organization
> of production, the system of material and moral incentives
> in work, and so on) and subjective factors (age, length of
> experience, occupational skills, and the like). [What takes
> place is that] *different demands* toward work arise—and

that the absence of real reliable knowledge of the *mechanism* of how these factors interact and of certain circumstances, to these demands (craving for creativity, development of skills and talents, and the like) not being taken into account in the practice of everyday [policy].[45]

3. *Technology underlies the main Soviet antidotes to alienation.*

Like other Soviet writings—and on the whole very much unlike Western writings on the same subject—*The Working Class and Technical Progress* speaks with fervor and optimism and, in some detail, about how technology now assures an end to alienation from work. On the basis of a socialist system, the reader is told, technology can be used in a wholly systematic way to shape and sustain the Soviet antidotes to alienation.

In brief, these antidotes revolve around three themes. One is collectivism, which here refers to the great subjective role of work teams for each worker who belongs to one.[46] Another key theme deals with varied and diversified work. Such work would fit in between narrow specialization and the ideal of eliminating the division of labor altogether. In the setting of work teams and full-blown collectivism, an all-around diversification of work comes as close to doing away with specialization as modern production allows.[47]

A third and last theme has been the subject of much Soviet research. It has to do with technical creativity, the "rationalizing" improvements that workers suggest for the equipment and procedures with which they work.

How large technology looms in the Soviet antidotes to alienation can be seen in a list of the main factors which, it is claimed, assure that a Soviet worker will have a "good" attitude toward his work. According to the Gorky

study, we find, each attitude can and should flow from
the new technology by:

> (a) taking into account the aim of shaping an integrated,
> harmoniously developed personality when new technology
> is built; (b) improvement of the organization and man-
> agement of production; (c) improvement in the conditions
> of work; (d) changes in the forms of interaction among
> people in the process of work; (e) [further] development
> of the system of material and moral incentives; (f) rise of
> each working person's initiative and responsibility in de-
> veloping production and the like.[48]

4. *While alienation from work reflects some contradictions within Soviet society, the main functional needs of
that type of society serve to do away with these contradictions.*

On the level of theory, *The Working Class and Technical Progress* assigns a notable place to a few concepts
that the new Soviet sociology, as a rule, does not stress.
These concepts are "contradictions" and "functional needs."

If one may generalize from these few comments, some
Soviet scholars now make two points. Alienation from
work can be found not only in capitalist societies but also
in modern societies without capitalism. In such a setting,
contradictions between creed and reality can and do arise
from changing technology, on the one hand, and, on the
other, from the make-up of individual personalities. These
individual personalities, the book notes, yield more slowly
to organized social change than does technology. Yet, like
earlier Soviet theory, this book does not dwell on the
basic issue of contradictions. While, in the words of an
American sociologist, Marx "remains the most powerful
analyst of asymmetrical relationships,"[49] the emerging
Soviet theory of work and alienation speaks in terms of
harmony rather than conflict. The concern here appears to

be not political conflict but economic efficiency. That is
not to say that conflict does not get its theoretical due.
But it does so only in an elusive form.

The constant Soviet stress on the role of work, and the
stated goal of making work fully humane and satisfying
in the near future, assumes a basic change in the class
structure. In both its theoretical sections and its data, tho
book on the Gorky study goes back to a central Soviet
theme: the social development of the Soviet system calls
for a steady fading away of the main division of labor
in modern societies—that between mental and physical
labor. In the Soviet setting, the functional needs of both
society and production call for an ever greater fusion of
intellectual and manual effort. With that, class divisions
must disappear, and hence also the fundamental source
of social conflict.

Thus, the functional needs of a socialist society rule
out contradictions such as alienation from work. All this
recalls much more the Western structural-functional theory of social equilibrium than the conflict theory of Marx.
How much this is the case can be seen from a concluding
passage on "social mechanisms":

> A social mechanism is an integrated, relatively stable sys-
> tem. It [consists of] a sum total of functionally linked
> material conditions and social relationships of people,
> which determine the norms of social interaction and be-
> havior of people in concrete social situations. The laws of
> the functioning of social mechanisms guarantee the gen-
> eral conditions of stable coexistence of people, and their
> organization within the framework of a single whole in
> the settling of specific tasks of production.[50]

These four propositions do not amount to any great
new theoretical construct. Yet it seems to me that, in the
context of this essay, the emerging theory of work and

alienation shows one thing: a formally established ideology need not preclude some scientific theorizing about
society. True, the theorizing I have sketched dealt with
only one layer of social organization and by and large
said nothing about the rest. Most important, the social
system as a whole and the realm of politics are treated as
givens. This suggests some limits to what social theory
can do in the Soviet setting. But as the last section of this
chapter argues, most theorizing in other kinds of sociology
also treats the entire social system as a given.

Piecemeal theorizing about major portions of social
reality amounts to something like the "theories of the
middle range" that command a good deal of support
among Western and especially American sociologists. In
each case, the value of such middle-range theories depends
on how well they can be tested empirically, and how well
they stand up when tested. As they were drawn here from
the Gorky study, the four theoretical propositions about
work and alienation do lend themselves to factual verification, now or in the near future.

There is a big gap, however, between current Soviet
theorizing and the Western notion of middle-range theories. Ideally, the latter rests on empirical data alone. In
the Soviet case, this new kind of sociological theorizing
calls for something else. A general theory of society—the
official body of Marxian and Leninist ideas—serves as
the point of departure. Not long ago, "concrete" research
tried to fit into it quite directly. Now a shift seems to be
taking place, both in the official scheme of social action
and in its view of the scheme of pure knowledge. In line
with this shift, a new "realistic" element in ideology and
the new "concrete" element in sociology may move in
harmony with each other. Together they would bring
about a distinct, Soviet type of middle-level theorizing.

This type of theorizing need not take issue with the

general theory. Its task would be to review, and to test by rigorous studies, a host of more tangible official ideas about social reality. The Gorky study, with its specialized theorizing about work and alienation, suggests such a development. So do recent pleas by two leading Soviet sociologists, Andreeva and Yadov.[51]

Since Stalin (and since Khrushchev), the Soviet state has faced a wide range of problems, old as well as new. To the extent that it will cope with these, we are apt to see strong support for both "realistic" ideology and "concrete" sociology—and close harmony between them.[52]

The Common Problem: Critical Analysis

In terms of the classic problem of science and ideology, this essay brings out two orientations in post-Stalin sociology, the future-oriented and the concrete. No clear line can be drawn between the scholars that follow one path or the other.[53] Nonetheless, some contrasts do appear. The future orientation shows a good deal less change from Stalin-type "total" ties between science and ideology. The Soviet sociologists who are now most active professionally tend to be closer to the concrete orientation. From them, too, come the more fully documented criticisms of Western sociology.

This raises a set of questions about the concrete orientation. At present, it seeks rigorous quantitative methods of studying reality and new data and generalizations about human interaction. Yet the question arises whether even "concrete" Soviet sociologists can move from limited descriptive surveys to work that aims at causal analysis. Can they move from studying individual responses to the environment to a structural analysis of the environment itself? And if they eventually do both, which is quite likely in the years ahead, can Soviet scholars ever bring any central part of the system into the scope of their investigations?

In many ways, of course, the answer is political. Within clear-cut limits, as noted, the Soviet state now seems to tie its future to a more subtle knowledge and handling of social reality. Hence it may well need to let a range of flowers bloom among the scholars close to it. Yet politics is only part of the story. The rest can probably be found within sociology. In this field, what kind of work can Soviet sociologists do?

Kurt H. Wolff has pointed up a deep gulf between "American" and "European" sociology—between individual-psychological realism (and social-historical nominalism), on the one hand, and very opposite metaphysical tendencies on the other.[54] In recent quantitative research by Soviet sociologists, we see much of the "American" individual realism and social nominalism to which Wolff refers. That does not mean that some Soviet sociologists are giving up their own view of science for the Western pluralist view. Rather, in our time the two views of science themselves may have developed a basic epistemological likeness.

As long as any science of society involves no methodically critical view of reality, sociologists using it are apt to treat the existing institutions and values as given. They look not at society as a whole but at lesser units within it, or at individual attitudes and adjustments to it. Today, this pattern applies not only to the Soviet kind of sociology but also to the main American kind. It could well be that only a systematically critical sociology can work against such reification of society.[55] Despite the very distinctive view they hold of science (and of ideology), Soviet sociologists now face the same major problem as do their colleagues in the West.

PHILOSOPHY

RICHARD T. DE GEORGE

Marxist-Leninist philosophy and ideology are closely interwoven in the Soviet Union. Soviet philosophy has a certain relative momentum of its own, and within limits the philosophical task of analyzing basic terms and beliefs develops according to its own internal logic. But Soviet leaders expect philosophers to render the ideology progressively more coherent and consistent, more believable and teachable.

The dual hat of philosopher and ideologist worn by the Soviet philosopher often has made him appear two-faced, if not two-headed. The introduction of conceptual clarity and rational consistency, and the philosophical inclination to follow an argument to its conclusion, often have tended within the Marxist-Leninist framework to pull in one direction, while the demands of supporting the prescribed ideological doctrine and the existing system have tended to tug in the opposite direction.

This chapter weighs the logical pushes and ideological pulls, and the resultant tensions of theory and the strains of practice, that have been operative in Soviet philosophy, and specifically in the Marxist-Leninist doctrine of historical materialism, during the past decade. Its purpose is both to provide a basis for evaluating the development of the doctrine and to elucidate the relationship between Soviet philosophy and ideology.

Philosophy and Ideology

As an ideology Marxism-Leninism contains a set of beliefs and values, derived from the Marxist-Leninist classics and interpreted by Party leaders and ideologists.[1] Marxism-Leninism serves two functions in the Soviet Union: (1) it influences and guides long-range action; (2) it is used to justify, or to attempt to justify, the action of the leaders of the Communist Party and of the Soviet government. Historical materialism as a fundamental part of the ideology shares in both these functions and so is not simply a philosophical theory of history or a neutral methodology of the historical or social sciences.

With respect to the first function of ideology, Marx, Engels, Lenin, Stalin, and the present Party leadership have consistently held that Marxism is not a dogma but a "guide to action."[2] From its inception it has evolved and changed. Its modifications have come always from above and most frequently under the pressure of practical necessity, in accordance with the Marxist doctrine that theory and practice are interrelated and reciprocally affect each other. To make sense of Marxism-Leninism as a guide to action, it is necessary to see it not as a frozen group of propositions from which deductions to specific actions are made, but as an adaptable and changing view of the world, a way of looking at it. This world view has inextricably bound up within it values and beliefs, many of which are not open to practical empirical verification. Within the ideology these beliefs and values are presented as factual or quasi-factual claims, though they do not actually function as such.

Such statements as "the triumph of communism is inevitable," "the aims of communism coincide with the aims of working mankind," "the Party as the vanguard of the working masses will lead mankind to communism," are not simply factual claims, though they seem to have the form of factual statements. As has often been demonstrated, these claims are future-oriented and are not open to direct empirical verification or falsification, at least not at the present time.[3] Because no factual evidence is sufficient to prove them, they cannot presently be held to express known, demonstrated truths. Rather, they express, first, a belief, and, second, a commitment to certain aims, aspirations, and values which a dedicated Communist is expected to help realize.[4]

Marxism-Leninism is a guide to action insofar as it presents the basic values and ends—the chief of which is the building of communism—that are to be fostered. In the realm of practical politics such ends and values supply only long-range objectives. They do not supply short-range practical courses of action, because a variety of specific actions are compatible with a given set of long-range goals. On the Soviet domestic scene they serve as a more proximate guide to the development of Soviet society, though here too they remain open to different particular applications and implementations. Marxism-Leninism as a guide to the action of *Party leaders,* more-over, is a more flexible guide than it is to the actions of simple Party members who have the more specific guide of the Party line to follow.

If Soviet ideology helps to guide Soviet practice, its more obvious function is to *justify* practice. Soviet leaders employ vast resources of propaganda to justify their ac-tions in the eyes of their own people (and thereafter in the eyes of whoever else cares to listen). They carry on openly and admittedly what they claim bourgeois leaders do surreptitiously—namely, they defend and justify their foreign and internal policies, the existing and developing Soviet institutions, the ends they hope to achieve, and the values they attempt to inculcate in their people. The money and time spent on producing mountains of propa-ganda literature, the extent to which the Soviet leaders go in propagating the Marxist-Leninist ideology, and the emphasis they place on ideological education in the U.S.S.R. are all sufficiently well known not to need be-laboring here.

What is flexible and changing in the ideology on the higher level is presented dogmatically on all lower levels. This is done precisely because on these levels it serves as the justification of action and policy initiated from above. It is only as a guide to action that Marxism-Leninism is flexible. Once a course of action has been decided upon and must be justified, the justification must be presented as fixed, true, and valid. Otherwise it would not in fact fulfill the justificatory function.

In contrast to ideology we might define philosophy as an attempt to understand the totality of reality on the basis of human experience and reason alone. To achieve its ends philosophers have traditionally engaged in two main types of endeavor: analysis of terms and basic beliefs, and synthesis of the results of the special sciences and common human experience. An ideology expresses beliefs, is generally uncritical with respect to its presuppositions, involves a commitment to a scheme of values and ends, and has as its purpose the guidance and justification of specific actions. Philosophy ideally deals with what is rationally demonstrable. It is critical, value free, and concerned with understanding rather than directly with action. Because ideological answers can be given to what have traditionally been considered philosophical questions, it is not always possible to distinguish clearly between philosophy and ideology. The two often are intermixed. They can best be distinguished not by the subjects with which they deal but by an analysis of their different aims and of their methods of approaching the questions they attempt to answer.

Within the Marxist-Leninist framework the task of Soviet philosophers is to explicate specific types of technical problems and to introduce order and clarity as far as is possible and consistent with the functions of ideology. The framework is dogmatic and uncriticized, and the philosophical function of clarification is constantly subordinated to the ideological functions. The basic framework consists of an interrelation of certain ends and values (communism is the chief end, collectivism a fundamental value), a particular methodology (the dialectical, historical method), a certain set of categories, and certain fundamental beliefs.[5] These elements remain constant and give Marxism-Leninism its continuity and identity. Built on these components and sometimes derived from them are certain propositions, some of them technical. These propositions have been variously interpreted during the development of Marxism-Leninism. They make up the technically philosophical aspect of the ideology.

Among Soviet theoreticians there are some who are

engaged primarily in philosophical analysis and clarification, though their analysis takes place within the dogmatic and hence ideological framework of Marxism-Leninism. We can distinguish two kinds of Soviet theorists. First, those who attempt to argue rationally for their positions, analyze and interpret the concepts they use, and attempt to reinterpret the meaning of the classical Marxist-Leninist statements in the light of rational consistency. Then there are those who merely repeat and paraphrase what is contained in the Marxist-Leninist classics and in the Party proclamations, who oppose any attempt at interpretation or reinterpretation from below, and who argue from authority.[6] In both kinds of thinker, however, there is the strain and tension of the demands of philosophy on the one hand and of ideology on the other. For philosophical clarity and consistency are pedagogically and so ideologically useful only if they do not undermine the ideological edifice itself. Yet safe, simple ideological conformity is often pedagogically and so ideologically sterile and ineffective. The stresses and strains are evident in the development during the past decade of the doctrine of historical materialism.

Three Levels of Historical Materialism

Historical materialism is "the science of the most general laws and of the motive forces of the development of human society."[7] It is a theory of history which asserts that the economic conditions, the productive forces and relations of any society are basic to that society. They determine all other social relations within it—for example, its legal and political institutions, its morality, art, philosophy, and religion. This theory is opposed to an "idealistic" view of history which holds that ideas and not economic conditions are the moving and determining force of history.

Historical materialism claims to describe both the general laws common to the development of all societies and the more restricted laws which apply only to certain societies. It also claims to present a method of studying social phenomena and to provide a means whereby the future development of society can be predicted and hastened. In its latter function it serves as a "guide to action." In this capacity and to the extent that it "expresses the interests of the working class," it is clearly and admittedly ideological.[8]

The claim that historical materialism is the only valid theory or philosophy of history has been challenged by critics for many reasons. Among these is the ambiguity of the terms in which the theory is set forth. The impossibility of historical foreknowledge of the type traditionally claimed by historical materialists has also been convincingly argued by many Western philosophers.[9] Such arguments and criticism have been rejected or refuted by defenders of the doctrine. They have had no discernible effect on the development of the position from Marx, through Lenin and Stalin, to the present. Yet in the past decade in the Soviet Union some serious controversies concerning historical materialism have broken out. Their resolution may well result in a considerably modified doctrine which may no longer be open to the same criticisms as heretofore.

The initial impetus for discussion came with the dethronement of Stalin in 1956, at the Twentieth Congress of the Communist Party of the Soviet Union. For the first time, it was then possible to question the correctness of the reinterpretations and innovations which had been introduced into historical materialism during the Stalin period. With the adoption of the Party Program in 1961 and the official proclamation that the Soviet Union had completed "building socialism" and was entering into the new phase of "building communism," a new urgency was given to the discussions, and new questions were raised. The result has been a number of debates and disputes concerning the status and presentation of historical materialism, its categories and laws, and the correct application and development of the theory as it applies to the Soviet Union itself.[10] Some of the debates are technical, theoretical, and aimed at strengthening an often ambiguous doctrine. Others have arisen from the practical necessity of guiding and justifying practice.

The changes in theory which have occurred have been related primarily to the development of Soviet society. The Soviet approach to developments in Western society shows no sign of changing. The world is still divided into sheep and goats; capitalism is still morally condemned, and its contradictions, it is claimed, will ultimately and inexorably lead to its downfall. This is not a hypothesis still to be tested, but a belief which is open only to confirmation, not to refutation. Thus the success of the Common Market does not (or did not) count as contrary evidence which might tend to disprove the future fall of capitalism, but de Gaulle's upsetting of the Western apple-cart does count as confirming evidence.[11] The belief seems comparable to the vague though fairly general belief that evil will reap its own reward and be its own undoing while good will win out in the end, rather than to the belief that Haley's comet will appear again in 1985. Marx's *Capital* and Lenin's *Imperialism* are sufficiently often cited and paraphrased so that the Soviet view of capitalist society can still be classified as "historicist" and is open to the arguments which have been brought against it on that count often enough and with sufficient force to need no repetition here.[12]

But according to Soviet theory, the contradictions in capitalism which in Russia led to the October Revolution were resolved thereby and thereafter, and the Soviet

Union has since been progressing from the realm of necessity where economic laws operate unknown and spontaneously to produce their effects, to the realm of ever increasing "freedom." Soviet society does not grow spontaneously but according to plan and direction, as spelled out in the Party Program, in five-year plans, and general Party directives. The practical necessity of attempting to pay off the promissory notes of Marx, Engels, and Lenin by delivering communism in the foreseeable future, however, has wrought changes of its own in the realm of historical materialism.

The dogmatic framework remains essentially the same. The demise of capitalism and the inevitable success of communism are predicted as certainties and presumably believed by Soviet leaders and theoreticians. Within this framework, however, is the realization that communism, as the Party Program emphasizes, must actually be *built*. There is both a belief that the laws of the development of communism can be found and a realization that the leaders of the Soviet Union do not have them. This in turn leads to the demand that particular empirical laws —economic, psychological, sociological—be found which can be used in constructive and specific social engineering and can be utilized as a basis for generalizing the laws of the development of communism.

We can thus distinguish three layers operative in the Soviet theory of society. The first level consists of the general laws of social development together with the predicted universal triumph of communism. This is the most dogmatic level and represents the original Marxist inheritance of the theory. The second level concerns the specific laws of the development of socialist society. This level is not yet completely solidified and is in the process of growth and formulation. It is the most dynamic of the three levels because it mediates the other two. The third

level comprises the study of the laws of particular sociological, economic, or psychological aspects of society. This
level is composed of the particular empirical disciplines
and the budding empirical research into particular facets
of Soviet society.

Within this scheme the second level supplies the goals
to be achieved by research on the third level. It is hoped
that these goals will be reached through the scientific
advances made by empirical research, technical advances,
and cybernetic techniques. Soviet theoreticians thus en-
visage that third-level research may enable Soviet leaders
to use piecemeal social engineering to replace the piecemeal improvisation which has characterized the development of Soviet society to such a large extent thus far. It
is further hoped that by achieving the goals of the second
level, the predictions of the first level will be fulfilled and
the theory as a whole vindicated. But one of the dangers
—if danger it be—is that the reinterpretation of the second order (so that the laws of socialist development serve
as aims or goals or directives rather than as descriptions
of necessary relations among social phenomena) tends
to infect the predictive certainty of the first-order laws
as well.

Within this tri-level structure of the theory of the
development of socialist society we can view in their
proper perspective the debates and discussions of the past
decade. They are of two kinds, the first stemming from
theoretical difficulties, the second from practical demands
of socialist development.

Disputes About General Theory

In some instances the need for clarifying the ambiguities
of the terms and laws of the general theory of historical
materialism became crucial for Soviet thinkers only when

the terms had to be applied to the development of Soviet society. Yet the theoretical disputes center primarily on the first level of theory. The terminological debates and controversies on theory have at least three causes: (1) the break between dialectical materialism ("diamat") and historical materialism ("histomat") which occurred under Stalin; (2) the inherent ambiguity of the terms and laws of the theory, as originally formulated; and (3) the development of new fields of learning and investigation. In each case the controversies are as yet unresolved, and every suggested solution has raised additional theoretical problems. Let us look briefly at each of the groups of controversy.

THE DIAMAT-HISTOMAT SPLIT

Marxist philosophy, according to Lenin's interpretation of Marxist doctrine, consists of dialectical and historical materialism. The former is concerned with the most general laws of the movement (dialectic) of reality (matter). These in turn are exemplified in the laws of historical and social development, which form the subject matter of historical materialism. But during Stalin's reign dialectical and historical materialism had undergone considerable modification from their classical presentation by Marx, Engels, and Lenin. In its Stalinist version the dialectic— the importance of which Stalin generally de-emphasized —became primarily a method of investigation. In the process one of the traditional dialectical laws (the law of the negation of the negation) was completely ignored.[13] At the same time the concept of "nonantagonistic contradictions" was introduced into Marxism-Leninism. The claim was made that antagonistic contradictions were no longer present in a socialist society and that consequently violent social changes—or dialectical explosions—were no longer possible therein.[14] The question of consistency that these and similar changes raised concerning the over-all theory of dialectical and historical materialism was largely ignored during Stalin's lifetime.

After Khrushchev's denunciation of Stalin, Soviet philosophers were quick to revert back to Engels' presentation of the laws of dialectics. In this domain the philosophical tug seemed strongest. But in historical materialism the ideological tug dominated. Here Stalin's innovations could not be repudiated without repudiating the theoretical basis and so the justification for many of the social developments which he had introduced in the Soviet Union. As a result, the Stalinist changes were accepted in some parts of Marxism-Leninism but not in others. Hence the rupture between dialectical and historical materialism, which had begun under Stalin, became more pronounced.

Theoreticians interested in dialectical materialism turned their attention to the difficulties raised by quantum mechanics, relativity theory, cybernetics, and theory of knowledge; those interested in historical materialism turned their attention to the laws of the development of socialist society. But each carried on without considering the implications of his theorizing for the other.

The resulting theories often pull in different directions. Thus, while the dialecticians attempt to clarify the claims that dialectical contradictions are the source of all development and motion and that changes from quantity to quality occur at "nodal" points, the historical materialists have adopted a quasi-mechanistic interpretation of the development of Soviet society. The latter argue that since the contradictions of Soviet society are nonantagonistic, its development is gradual. Whatever "leaps" occur take place gradually and over extended periods of time. This position seems very close to the nondialectical, mechanistic interpretation of the development of society championed by some Soviet philosophers in the 'twenties. Yet it must be reconciled with the claim of the dialecticians that the three laws of dialectics hold universally. If, as some argue, the process of socialist development is truly a new dialectical process, its relation to the general laws of dialectics must be clarified. If, on the other hand, the process is not radically different from the development of capitalist society, the claimed impossibility of explosive leaps (or of another violent social revolution) must be explained.[15]

The split between those interested in dialectical materialism and those working in historical materialism became clearer as each group progressed with its own investigations and clarifications. In the realm of the theory of knowledge, analysis has led some Soviet philosophers to hold that mind is a property of matter as such.[16] It is present analogously or potentially in all of matter, and the reflection by mind of matter takes place in some analogous way on all levels of matter. In this way they claim to be able to account for the origin and development of human consciousness. But this solution to one philosophical puzzle, whatever its worth, raises questions concerning the doctrine of historical materialism that social consciousness reflects social being. Two choices exist. Either the relation between social being and social consciousness is completely parallel to that between matter and mind, and so social consciousness must exist in some analogous way in social being (a position no one has yet openly espoused), or the two sets of correlates are not as parallel as had been previously held.

The point here is not that the necessary theoretical accommodations within Marxism-Leninism are impossible. All these questions are still being disputed in the Soviet Union and solutions may be forthcoming. Rather, the point is that changes in historical materialism often require changes in dialectical materialism and vice versa, one change in theory requiring another as a result of theoretical and logical considerations alone and irrespective of practical needs. Recent discussions[17] show that Soviet philosophers are at last conscious of the rupture in Marxism-Leninism between historical and dialectical materialism and that they are concerned with repairing it. It is now admitted that the categories or basic terms of dialectical materialism must be correlated with those of historical materialism, and an attempt is under way to rejoin these two elements of Soviet philosophy into a viable unity. How well Soviet philosophers will succeed in doing this remains to be seen. But their concern leads directly, and adds urgency, to the present attempts at clarifying the terms and laws of historical materialism, especially on the first level of theory.

THE AMBIGUITY OF THE ORIGINAL DOCTRINE

The materialist conception of history was never adequately or unambiguously stated by Marx or Engels, and ambiguity has plagued it since its first formulation. Stalin, in his pronouncements on linguistics, attempted to clarify the meanings of the key terms, base and superstructure, and in the process opened up a whole new area of claiming that some phenomena—language in particular—fall neither within the base (the economic substructure) of society nor within the social superstructure. Since no intermediary realm exists, language falls outside of the base-superstructure dichotomy.[18] This attempt at clarifying some of the basic terms of historical materialism has been continued and is presently being systematically pursued by some Soviet philosophers. The need for clarification has led to the analysis of basic terms or categories used in stating the theory,[19] the analysis of the laws enunciated by the theory,[20] and an investigation of the relationship between categories and laws.

The ground being trod here is delicate because the

categories and laws are central to historical materialism and any tampering with them may essentially change the nature of the doctrine. But even for the Soviet theoreticians to admit that there are difficulties, that their doctrine needs clarification, and that the base-superstructure relation—central to the doctrine—is in reality very complex, is an advance over their previous self satisfied posture.

The base-superstructure relation is still being discussed, and there is as yet no agreement about which social phenomena belong to the base, which to the superstructure, which to neither, and which to all three of the groups mentioned.[21] Attempts to clarify the meaning of "base" have led to distinguishing the "economic base" from the "technical base" of society.[22] But these distinctions, made in the hope of adding precision to the theory, have led to the unresolved question of the relation of these terms to one another and to the various elements of the superstructure. Moreover, once the base-superstructure relation was questioned or challenged, further questions and difficulties arose.

Thus, the meaning of "social being" and "social consciousness" has been similarly questioned. One group, in which V. P. Tugarinov is most prominent, holds that "social being" includes all the productive and social activity of men, including many superstructural relations (such as political institutions), and that its opposite, "social consciousness," is limited to the spiritual or intellectual activity of men, including ideas about such practical activities as production.[23] This approach gains force because it parallels the relation of the terms "being" and "consciousness" in dialectical materialism.

Another group—represented by Grigori E. Glezerman —denies the parallel and plausibly claims that "social being," since it refers to an aspect of human society, necessarily includes elements of consciousness. For this group "social being" includes those human relations which arise necessarily in the process of production and play a role in determining other elements of consciousness; thus "social being" and "social consciousness" are not in absolute opposition, and some relations, such as political relations, cannot be placed wholly in either category.[24] Similarly, the exact meaning of "productive forces," "means of labor," "instruments of labor," and so on, is disputed.[25]

The attempt at analysis and clarity here is typical of the philosophical endeavor. This is an area where philosophical techniques of analysis are being utilized within the Marxist-Leninist ideological framework. That analysis in turn has led to discussions concerning the meaning of the laws of historical materialism which utilize these terms, as well as to general discussions of the relation of categories and laws. Some argue that the categories are more basic than the laws in which they are used, and that changes in the meanings of the terms result in changes in the meaning of the laws. Others argue that the laws of historical materialism have already been proven, are true and thus primary, and must be used to help clarify the meaning of the categories.

At issue are such basic laws of historical materialism as "the determining role of social being in relation to social consciousness," "the determining role of the base in relation to the superstructure," and "the determining role of the means of production in relation to the social structure of society," all of which are central to the materialistic conception of history.[26] The precise meaning and extent of the active role of the superstructure in the development of society—another theme brought to prominence by Stalin and continued by his successors because of the necessity of justifying the direction of society by the Party—raise further serious questions concerning the meaning of "determination" in these laws. In all these cases the fact that the key terms operate on both the first and second levels of theory has made a solution particularly difficult.

The discussions which have developed are thus not on peripheral areas but come close to the heart of the ideological doctrine, though thus far all attempts at clarification have had as their avowed purpose the strengthening of the doctrine by making it clearer and more rationally consistent and coherent. The resultant unresolved ambiguities, differences, uncertainties, and difficulties may make the doctrine itself suspect and doubtful in the eyes of some. All these discussions, however, take place within the pages of philosophical journals and in books published in small quantity, while the official instruments of education and propaganda give no hint of the controversies. The philosophical controversies are thus kept within limits and are not allowed to impede the justificatory function of the ideology for the masses.

THE DEVELOPMENT OF NEW FIELDS OF LEARNING

Another set of disputes and debates has more practical consequences. These arose as a result of the growth of knowledge and the scope of investigation within the general domain of historical materialism. All questions relating to the development of society were traditionally considered in the domain of historical materialism. It provided both the method and the context for the investigation of social development in its various spheres. And while historical materialists admitted that there are certain regularities peculiar to specific areas of life, they enunciated the general laws of social development and rested content with this. A case in point is the domain of morality.

Historical materialism teaches that morality is a form

of social consciousness, that it arises as customs arise, and that changes in the economic conditions of society produce consequent changes in social customs, and so in morality. The dependence of morality on the economic conditions of society had been clearly stated by Marx and Engels, and the official Marxist-Leninist position on morality long seemed to indicate that the question of morality had been adequately handled in the classical Marxist texts.

For a variety of reasons, however, ethics, or the science of morals, began to develop in the Soviet Union early in the 1950s, and the philosophical literature concerning ethics and morality has been growing at a rapid pace since 1956.[27] It was recognized that moral as well as material incentives were essential to increase labor productivity, and that a new morality had not followed upon the change in the economic structure of the Soviet Union and the introduction of new productive relations. For although economic relations are supposed to determine social relations in the *final* analysis, Marxist-Leninist theoreticians (and perhaps also the Soviet leaders) have come to realize that there are a great many intervening links.

Thus, morality and other domains of social life were admitted to have a certain "relative independence."[28] The morality of a given time was seen as being not only the result of the economic conditions of a society but also a product of the society's past customs and morality, its former ideas and values, the psychological make-up of the people, their level of knowledge, and other similar factors. The investigation of changes in morality therefore required a specific study, and ethics has become an independent branch of knowledge. This in turn has raised such questions as: Where does historical materialism end and the independent study of ethics begin? What is the relationship between the two? A similar situation has developed with respect to sociology, social psychology, and "scientific communism."[29]

Here all three levels of the theory of historical materialism come into play. The first-level theory claimed that changes in productive relations would produce changes in the superstructural social relations. But on the level of socialist society it became clear that socialist relations of production had not wrought significant changes in Soviet society and had not succeeded in producing the new man—whether the cause be blamed on the lag of consciousness behind being or on something else. This in turn has emphasized the need for third-order empirical research, sociological and psychological as well as economic. One result has been increased emphasis on sociology as an empirical discipline, which is now breaking away from historical materialism in the Soviet Union as an independent branch of study.[30] But this has raised problems concerning the relationship of historical materialism and sociology, both for pedagogical purposes and for practical approaches to empirical investigation.

Historical materialism had long been identified with Soviet sociology because it studied the laws of social development. But with the development of an empirical, specialized discipline, the status and organization of historical materialism became open to question. More important than the terminological question is that of the actual reciprocal relation of third-level empirical sociology and the other two levels of historical materialism. F. V. Konstantinov, a leading Soviet ideologist and philosopher, and V. Zh. Kelle, his colleague at the Institute of Philosophy in Moscow, claim in an article in *Kommunist* that "the needs of development of Marxist sociology, i.e., of historical materialism, demand the wide application of concrete sociological investigation."[31] They add that the Party should help both to develop sociology and to make wider use of its findings.

Similarly, D. M. Gvishiani, writing in Voprosy filosofii, underlines the necessity of uniting philosophy with practical activity and of developing historical materialism by utilizing the findings of the other social sciences. "Any investigation in the domain of historical materialism," he continues, "can and must be concrete, whatever general questions it is concerned with. Otherwise it simply is not scientific investigation."[32] D. I. Chesnokov, a leading elder expositor of historical materialism, has also emphasized the need for sociological investigations into the "socioeconomic formation of the contemporary epoch" in order to serve as a basis for developing the laws of historical materialism.[33] Even Leonid F. Il'ichev, who, under Khrushchev, was the official Soviet ideological spokesman, claimed that there could be no philosophical understanding of the laws of social life without concrete social investigation.[34]

Exactly how these statements are to be taken is not clear. The Party is certainly being encouraged to utilize the results of empirical research to manage the country more efficiently. But whether this third-level research is expected to affect the second level is still a moot question. The need for finding the laws which govern socialist society is pressing. If the incipient empiricalization of the third level were allowed to penetrate into the methodology of the second level, it might have very far-reaching effects in changing the present a priori and dogmatic structure of historical materialism. But whether historical materialists are simply to generalize on the basis of empirical economic, historical, and sociological findings, as Chesnokov seems to suggest,[35] or whether such generalizations are themselves to be tested empirically, as A. M. Gendin seems to suggest,[36] has not yet been decided. The latter would be a more far-reaching change than the former; but even the simple utilization of data from empirical investigations would be a step forward in the development of theory. For one of the shortcomings of historical materialism has always been its failure to account for all the pertinent empirical data and its tendency to generalize too quickly and too superficially on the basis of rather vague laws of determination.

The Philosophical Problems of "Building Communism"

If the need for clarity and consistency is bringing about certain changes in theory, then practical necessity is doing so even more. Two different kinds of practical difficulties have led to changes in theory. The first is the failure of changes in Soviet economic structure to bring about the desired and expected effect in Soviet social relations and in the Soviet people. The second difficulty stems from the fact that no one has successfully uncovered a significant number of "laws of the development of communism" of sufficient precision to be of much help in specific planning. The changes which have resulted in historical materialism consist primarily of the development of the second level of theory. The compatibility of the second-level developments with the first level of theory, however, may well be questioned. These developments, moreover, seem to be significantly infecting the first level. Let us first clarify the nature of the practical difficulties and then examine the tensions which have resulted between the first and second levels.

THE SOVIET BASE AND SUPERSTRUCTURE

By the 1950s, Soviet theorists argued, capitalism had been replaced by socialism in the U.S.S.R., private ownership of the means of production had been done away with, and the economic exploitation of one individual by another had been eliminated. Yet it was quite obvious that the Soviet people and the structure of Soviet life had not changed drastically. New social relations did not develop with new economic relations, the laws of historical materialism notwithstanding. If the new communist society really required "new men" to build and people it, it was obvious that they must be formed and that they were not being formed automatically. As a result, the theme of "the active role of the superstructure" in social development came openly to the fore as guide and justification of necessary practice. It has gained added momentum as a result of the 1961 Party Program. If new men were not formed as a result of changing economic relations, then the new *homo Sovieticus* must be molded, his ideas and values changed directly by other means.

The concept of the active role of the superstructure was present in the doctrine of historical materialism from the time of Engels, for he did admit that the superstructure influences the base. But for him it was certainly secondary and subsidiary. In the hands of Stalin the superstructure in some instances took the form of a "motive force" in society[37]—a term reserved in classical Marxism for the economic conditions of society. For, to admit that the superstructure—ideas or theories, morality or patriotism—is a moving force in society tends to erase the line between the materialist and the idealist interpretation of history. The Stalinist line in this respect has been continued, though Soviet theoreticians are quick to add that it is essentially and ultimately the economic conditions that are the determining factor in history and social development. Such disclaimers, however, do not mitigate the change in emphasis in the original theory, and the change itself encourages further clarification of the terms "base" and "superstructure" and a more detailed analysis of their interrelation.

The theme of the active role of the superstructure supports and justifies the continuing direction of society by the Communist Party. This is nothing new. But the Party directs society not only by directing changes in the economic structure but also by acting directly on the people. For the base to develop according to plan, the workers must be motivated to work with ever more diligence for the building of communism, otherwise that edifice will never rise. Thus, the theme justifies the use of moral force to aid in the development of society. The Soviet worker is to be motivated by moral as well as by material incentives to work for communism; morality is to be used to increase productivity by promoting initiative, diligence at work, care for the tools of labor, and so on.[38] It is to replace physical force, violence, and terror as a means of social control.

The theme of the active role of the superstructure supports the Soviet attempt to change the social consciousness of the Soviet people—their ways of thinking and looking at things, their value schemes and customs—not indirectly by changing their economic life, but directly by education, propaganda, and ideological indoctrination. This is both the rationale and the reason for the stepped-up ideological campaign that followed the adoption of the 1961 Party Program.

The need for such inculcation has increased emphasis on the need for the development of social and educational psychology. The aim is to discover how ideas and values can be inculcated most effectively, how old values can most effectively be replaced by new ones, and how moral stimuli can most effectively be used to increase productivity. Here, then, is a specific case in which ideological needs have spurred empirical research, not so that these results can be used in changing theory but so that they can be utilized in more efficient inculcation. The need for successful inculcation becomes clear if the new man is to be formed, if the new society is even to be approached, and if nonviolent means of social control are to replace the rule of force. The Soviet attempts to change man directly run up against the difficulty of such disvalues as oppression, alienation, and the division of labor. Socialism was to do away with these. But they are still distressingly present in Soviet society. And there is no clear plan of how they are to be eliminated in real life, as against just verbally denying that they exist. But present discussions of these topics seem to indicate merely that the meaning of the terms will be reinterpreted so that they can apply only in a class society.[39]

THE FAILURE TO DISCOVER SIGNIFICANT LAWS OF SOCIALIST DEVELOPMENT

The change wrought in theory by the fact that adequate laws of the development of communism have not been found is more serious. It is central to the heart of historical materialism. The laws of the development of capitalism and its consequent demise had been outlined by Marx; but neither he nor Lenin had studied the laws of the development of socialism, as it did not exist in their time. Neither a Marx nor a Lenin has appeared on the Soviet scene with the laws necessary to guide practice or to be used by the leaders in guiding Soviet society in any detail. The general laws of the development of all societies which historical materialism can supply consist of such general claims as the determining role of productive forces in relation to productive relations; the determining role of the means of production in relation to the social structure of society; the determining role of social being in relation to social consciousness; the determining role of the base

in relation to the superstructure, and so on—all of which
are undergoing reinterpretation and in all events are too
vague to serve as guides for specific action in the development of socialism into communism.

Soviet society, it is claimed, will increasingly enter the
domain of freedom the more it knows the laws operating
within its economic structure and can utilize them to
guide the society to the ultimate construction of communism, which is to be achieved "in the main" by 1980.
But it is all too clear that knowledge of such laws is
lacking.

If we look at the laws peculiar to the development of
socialism which have thus far been incorporated into historical materialism, we see that they do not describe
necessary relations between social phenomena, as the
general laws of social development do when they speak
of the "determining role" of one part of society on another. According to the recent *Philosophical Encyclopedia*,
some of the laws peculiar to socialism are: the planned
and proportional character of economic development, the
absence of crisis production, the relation of collaboration
and mutual help among men, and the absence of antagonistic classes.[40] These "laws" are either *tautologies*, giving
a definition of "socialism" such that unless the conditions
mentioned are fulfilled one cannot properly say that
socialism is present, or *guides* for the proper development
of socialism.

Thus the "law of the absence of antagonistic classes,"
which says roughly that there are no antagonistic classes
in a socialist society, is stating a stipulative criterion by
which to judge whether or not socialism is present. If
one finds antagonistic classes in a society, then that society cannot properly be termed "socialistic," for the absence of such classes is a necessary though not sufficient
condition of a socialist society. The "law of the relations

of collaboration and mutual help among men" states that such relations are necessary to the development of socialism and (perhaps) that they are fostered by the development of socialism. Because these relations still exist in the Soviet Union on too small a scale, the Party—through its new moral code and by other means—is attempting to foster such relations among the people. Here the law states not what is the case, not a necessary relation of fact, but what the Party desires to bring about. "The absence of crisis production" states an aim which the leaders of Soviet society hope to achieve through planning, and which they believe comes about as a result of proper utilization of economic laws. But this law too seems to be arrived at a priori by the simple analysis of either the meaning of the term "socialism" or the aims of the leaders of the Soviet Union.

A large difference exists between the laws enunciated by the founders of the materialist conception of history in describing general social development of all societies and the laws enunciated more recently and applicable to socialist society. This difference seems characteristic of a perhaps unnoticed change which has taken place within the framework of the classical doctrine of historical materialism.

A New Determinism?

On the first, most general level of theory communism is still held to be inevitable—evidently in some strong sense of "inevitability," meaning that it will certainly come about necessarily and inexorably. This has been and continues to be a strong component of the ideology. It justifies action done in the name of hastening the inevitable, assures followers that they are on the winning side, and puts them in the line of progress making them inheritors of the great future. As far as promoting specific action, however, or directing specific empirical research, such inevitability is of little help.

The perhaps unconscious reinterpretation of historical materialism on the second level, the level of finding the laws of socialist development, consists of a shift from the factual and declarative aspect of inevitability to the emotive and hortatory aspects which have always been present. This subtle reinterpretation takes the inevitability out of the word "inevitable," the necessity out of the word "determined," and the universality out of the word "law." We see this if we go not to the textbooks of historical materialism but to the journal articles where the current discussions are taking place. There we find the vocabulary of historical materialism has not changed; yet an analysis of the way in which the terms are used on the secondary level is a clear indication that they no longer have the same meaning or serve the same function as on the primary level.

Thus G. E. Glezerman, in a discussion of the dialectic of objective conditions and subjective factors in the building of communism, emphasizes that "building communism" has no precedent in history and consequently it would be naive to think that it can be achieved without a search, without trying different forms of organization and various social experiments. He suggests experimentation on a small scale to test varying solutions to a problem before making any changes on a country-wide level, and he advocates social experimentation in schools, kolkhozes, and economic districts.[41] Thus he admits that knowledge of the laws needed for directing Soviet society to its desired end is presently lacking and he underlines the need to find these laws. The inevitability of communism is here definitely secondary to the directive concern. The logical corollary, namely, that the needed laws might never be

found and that communism might not be achieved, however, would undoubtedly be rejected by the author in
the face of a prior first-order ideological commitment.

In discussing the problem of values, I. F. Balakina
states that man not only knows the world as it is but
also subordinates it to his subjective ends, thus implying
that the world can be molded in different ways.[42]

Piotr N. Fedoseev, writing in *Kommunist*, underscores
the growth of the "subjective factor" in social life, evident in the increasing role of consciousness in the direction of society and in the necessity of Party guidance.[43]
But the growth of the "subjective factor," a developing
theme in the literature of historical materialism, sits somewhat uneasily in a theory of inevitability, which previously
seemed to speak in terms that made psychological choices
and motivations secondary to more basic processes of
historical determination. M. N. Rutkevich, addressing himself to the concept of development, distinguishes between
progressive and regressive development. He claims that
there is no definite direction in the cosmos and he reduces
the claim of the inevitability of the triumph of communism
to the much milder and more plausible claim that socialism offers the possibility of progressive development.[44]

F. V. Konstantinov and V. Zh. Kelle reject a fatalistic,
Laplacian, mechanical determinism and point out that
any existing social conditions offer different possibilities
of development and that which of these possibilities is
realized depends on a variety of factors, not least of which
is what people decide to do. Though they speak of world-
wide socialism as being inevitable, its inevitability is not
categorical but contingent on many factors.[45] Fedoseev
accordingly differentiates between "automatic laws" which
operate necessarily and "action laws" which are not automatic but require planning.[46] Thus the proportional distribution of all the resources of society so as to produce

the rapid growth of productive forces and the maximal satisfaction of citizens' needs is an "action law" of socialism. V. N. Cherkovets similarly claims that while the laws of capitalism operated spontaneously, there is a qualitative difference between them and the laws of socialism in which the conscious element becomes an indispensable feature of their operation.[47]

These and similar statements provide the basis for an interpretation of the Soviet "science of *socialist* society": in it a law often becomes equivalent to a trend or an aim or a desired correlation. The subjective factor in these laws is twofold: first, conscious intervention and manipulation is required in their use; second, the manipulation is toward the conscious achievement of some goal or end or value. Such laws, which are part of the second level of theory, operate within the first-level framework. But the two different meanings of "inevitable" found on the two levels cause a tension which has yet to be either noticed or faced by Soviet theoreticians. To give up the strong meaning of the term would involve serious revision of the entire doctrine and would undercut much of its ideological force. But to insist on the strong meaning of the term on the second level of theory would be to hamper its directive function and the search for the empirical laws necessary to enable Soviet society to reach communism.

If conscious, goal-directed activity is necessary for producing communism, then the term "inevitable" with respect to socialism and communism is reducible to the claim that socialism alone offers the possibility of realizing and implementing communist values and ideals. The concept of determinism as used by Soviet theoreticians is here no longer categorical but is shot through with contingency; certain results will be achieved *if* there is no world-wide nuclear war, *if* the proper means are found, *if* the proper decisions are made and the proper action taken, and so forth.

The Soviet claim that their theory is scientific, as opposed to being an instance of "voluntarism," subjectivism, or utopian daydreaming, is reducible to one claim: that their goals and aims—both long-range and proximate—are based not merely on their subjective desires but on knowledge of the objective conditions necessary for their fulfillment. Though the jargon remains, what these phrases mean turns out upon analysis to be platitudes which almost everyone would accept—namely, that given certain objective conditions various alternative activities are possible, while others are precluded. The only test of whether an action is really possible is practice; success separates voluntarism from science.

Soviet theoreticians attempt to tread a narrow line between fatalism and idealistic voluntarism. But if choice, desire, aim, and will are to play as important a role in Soviet society as present Soviet theory maintains, then the claim of historical inevitability must be softened throughout the theory to make room for it. The Soviet ideal is still total social control and planning. Yet practical exigencies have forced the at least implicit admission that many of the laws necessary for holistic planning are missing and that experimentation and improvisation are needed until additional laws are found.

The Changing Role of Values

Under these conditions the actual guide to Soviet social development is a set of values, though it is stated in the form of laws. The ultimate value to be achieved is communism. Such subsidiary values as mutual helpfulness and collectivism, and such conditions as the absence of classes and of crisis production, are seen as leading to the desired end. Soviet society has progressed sufficiently, however, so that it is becoming more and more difficult and expensive in terms of time and money to see just how further progress is to proceed, how best to match and adjust material goods and individual and social needs, while at the same time producing and cultivating social, cultural, moral, legal and aesthetic values. Because of the vagueness of the ideal of communism, other proximate norms for decision-making are being used, though practical necessity often forces judgments without sufficient knowledge or consideration of the objective value involved and on an *ad hoc* basis. The task of a Marxist-Leninist value theory, according to one philosopher, is to investigate objective natural and social regularities so these can be used to promote the progressive development of society.[48] To date, the most detailed scheme of values, as well as the most detailed plan for the building of the new communist society, is the Party Program, which speaks of the "rational needs" of the people being satisfied in the not too distant future. What these rational needs are will presumably be decided by the Party leaders and planners—not by the philosophers; on what basis they will decide it is difficult to say. But the need for developing a comprehensive scheme and theory of values, though completely ignored until the late 1950s, is now generally admitted.[49]

On the interpretation of the laws of socialist development which sees them as stating values and ends to be realized, subjectivism remains a threat. The newly developing Soviet theory of value being worked out by Soviet philosophers accordingly distinguishes subjective values which reflect individual or collective desires from objective values which represent the real needs of the development of society.[50] It is the objective values and ends of society, not arbitrary or subjective values, which the laws of socialist development attempt to present.

What I am suggesting is twofold: first, that practical necessity has forced Soviet ideologists to begin to do what all the anti-Marxist critiques of the Marxist interpretation of history could not force them to do. It has forced them to analyze and revise some of their prophetic claims and to start on an empirical study of their own developing society. This empiricalization both bolsters the hope of achieving communism and, to a limited and still controlled extent, threatens the dogmatic framework within which it operates. Second, Soviet society is to a large extent being guided in its development not by laws of social development but by a certain set of values or ends or ideals, which are part of its Marxist inheritance and give it more than verbal continuity with historical Marxism.

If this is the case, then there is no a priori reason for holding that the Soviet leaders cannot succeed in molding their society to their own image, at least within certain tolerable limits.

An effective critique of historical materialism as it is developing should go beyond simply exposing it as historicist or unscientific or muddled, with the consequent implication that it can be dismissed. It presents a model of what Soviet society—or at least its leaders—are trying to achieve, an ideal self-portrait of its future which it hopes to construct, a forecast or prophecy, the very existence of which may help its achievement. Models are not true or false, but only more or less adequate, more or less realistic as guides to action. An adequate critique should therefore also analyze, explore, and if necessary expose the values which historical materialism promulgates and which the Soviet leaders are attempting to inculcate in the Soviet people. The logic which is pertinent

is not only that of science or of history. It is also that of
values. This involves the relation of values and facts, of
commitment to values and the performance of certain
actions, of the compatibility or incompatibility of the simultaneous development of certain values.

There is no doubt that, as a result of the factors just
outlined, historical materialism is in something of a state
of flux. But how seriously the present discussions will
affect the theory as a whole, and what the significance of
any resulting change in theory will be, are both moot
points.

Analysis of the basic categories of historical materialism and of its laws could lead to a drastic revision and
undermining of its claims. An example of how a thorough-
going analysis of the categories and laws of historical
materialism can discredit the theory is supplied by Poland.
There Marxism-Leninism as a whole was subjected to such
analysis and criticism by Polish philosophers trained in
logic and analysis. As a result of their endeavors, Marxist-
Leninist philosophy—which for a time was the only philosophy officially recognized or tolerated—is now one of
several philosophical streams in Poland,[51] though it remains the official philosophy and as such is a component
part of Polish Marxist-Leninist ideology. But lest the
Polish model be taken too seriously, it is necessary to
remember two things. First, the Soviet Union does not
have the tradition of logic and analysis which existed in
Poland. Second, the Soviet analysis which is taking place
is being done not by outsiders who hope to challenge the
theory but by dedicated Marxist-Leninists who hope to
strengthen it.

The Soviet work of clarification is taking place within
the Marxist-Leninist bounds. There is no reason to believe
either that Party leaders will allow any changes in the

theory which might threaten to undermine the ideology or its power or that any Soviet philosophers wish to do so. It is too early to tell how successful the philosophers will be in making historical materialism more cogent, consistent, clear, and convincing than it has been up until the present. But no matter how limited their success, it seems that any success will strengthen and not weaken the system. An unexpected degree of success and agreement in resolving the present disputes and controversies may well make the position much more teachable, believable, and acceptable, even to skeptical intellectuals.

The trend toward empirical research in the attempt to find the "laws of the development of communism," and the tendency to adopt the results of empirical research in other areas, might be considered a step toward the continuing erosion of ideology and its replacement by a pragmatic approach to specific problems. But here too the case on the other side seems at least as strong, if not stronger. For the empirical research itself operates and functions within the ideological system. What is studied and the aims sought are dictated at least in part by the ideology. Thus empirical research can ascertain where the Party is failing, where indoctrination is not successful, where the people are dissatisfied; then the proper means can be taken to remedy these deficiencies.

Marxism-Leninism as a body of doctrine has within limits changed over the years and continues to change. Such change, however, does not necessarily spell its demise, but it may be considered an index of its strength and flexibility. Several elements remain constant: a dedication to the end of communism and to a set of values related to its achievement; a distinctive approach to problems which can be called its methodology; a certain set of categories in which the world is conceived; and a certain minimal set of propositions which are not presently being questioned or challenged. Recent developments in the field of historical materialism, therefore, seem inclined to make it both more systematic and conceptually consistent and more closely related to practice, more realistic. As such these changes can be considered progressive, according to Soviet theory, and will tend to strengthen rather than weaken the ideology of Marxism-Leninism.

Philosophers in the Soviet Union today, it seems, may help reinterpret the world and improve the official doctrine on its secondary and tertiary levels. Yet in the foreseeable future their efforts will probably not change the basic Soviet ideological framework in any significant way.

CYBERNETICS

LOREN R. GRAHAM

Cybernetics today enjoys more prestige in the Soviet Union than in any other country in the world. This observation may seem surprising to Westerners, particularly to those who know that the field was at first severely criticized in the Soviet Union. Furthermore, the Soviet Union lags behind the United States in computer production, both quantitatively and qualitatively. How then can Soviet writers constantly speak of the unique role which cybernetics will play in their society? The following chapter attempts to answer this question by analyzing the essential concepts of cybernetics against the background of traditional Soviet aspirations and ideological commitments.

Cybernetics, at first glance a narrow topic, turns out to be intertwined with issues relating to both man and physical nature. Any development which causes reconsideration of such basic issues inevitably touches upon dialectical materialism, which is described in the Soviet Union as the science of the most general laws of nature and thought. Furthermore, cybernetics may be applied toward the achievement of social goals, and thus affects Soviet progress toward communism, the final stage of development according to historical materialism. In sum, the discussion of cybernetics is clearly a new and interesting example of the constant interplay of science and ideology in the Soviet Union.

This latest discussion bears some resemblances to past Soviet debates in science, such as those over biology and physiology, but it is distinguished by several totally new characteristics. The most prominent of these is that the cybernetics controversy has resulted in an overt effort to develop the field more rapidly, in contrast to past suppressions or retardations of particular sciences following ideological discussions. Western analysts are thus led to consider the question of whether the relationship between science and ideology is not much more subtle than previously thought. Is a conflict between science and ideology inevitable? Or are there also agreements between certain aspects of ideology and certain scientific developments?

The Soviet Striving for Rationality

An original promise of the Russian Revolution, for those who supported it, was the rational direction of society. Marxism as an intellectual scheme was heir to the optimism of the French Enlightenment and the scientism of the nineteenth century; one of its primary characteristics was the belief that the problems of society could be solved by man. Nature was not so complicated but that it could be controlled if only the artificial economic barriers to that control erected by capitalism were removed.

The key to progress, then, according to the Marxists, was social reorganization. The Bolsheviks considered the Rovolution of 1017 to be the decisive breakthrough toward that reorganization. They admitted, of course, that progress toward efficient administration would be very difficult to achieve in Russia as a result of its primitive state. Even in the early years of Soviet Russia, however, there were at least a few theorists who hoped to achieve centralized, rational direction. The first attempt toward this goal was made during the period of War Communism (1918–1921). However important the Civil War may have been in forcing a command economy, it is quite clear that the ideological urge to create a planned, communist society also played an important role. From this standpoint the New Economic Policy (1921–1927), with its relaxation of economic controls, was a definite retreat. The rapid industrialization that succeeded the New Economic Policy might have been carried out in accordance with any one of several different variants, but all assumed greater planning and centralization.

After the 1930s, however, the goal of a rationally-directed society became more remote. The most disheartening fact to the Soviet planners was that the more the early difficulties of industrial underdevelopment were overcome, the more distant seemed the goal of rational, centralized control. By the time of Stalin's death in 1953 the economy had become so complex that it seemed to defy man's ability to master and plan it. It would have been convenient to attribute these troubles to the irrationalities of Stalin himself rather than to the inability of Soviet man to control his affairs. Yet by 1957, four years after Stalin's death, it was clear that the trouble lay not in the aberrations of one man but in the entire concept of centralized planning.

Notwithstanding the rhetoric which surrounded the reform, the decentralization of industry that occurred in 1957 was again a defeat for rationality—at least as rationality had originally been conceived in the Soviet Union. There were very serious grounds for the belief that a complex modern industrial economy simply cannot be centrally directed. Every modification of the quantity of one commodity to be produced called for unending modifications in the quantities of others. Even a relatively decentralized economy seemed to have an insatiable demand for bookkeepers and administrators. Academician Glushkov said that if things continued as they were going, by 1980 the entire Soviet working population would be engaged in the planning and administrative process. To use a cybernetic term, the entropy of the system was multiplying at a horrifying rate.

It was at this time in the history of the Soviet Union that cybernetics appeared. Leaving aside temporarily the initial Soviet hostility toward cybernetics (which has been exaggerated in the West), the promise of cybernetics, as it appeared to Soviet administrators and economists, was twofold: first, it held out the hope of rational control of processes which previously had been reluctantly judged uncontrollable because of their complexity; second, it offered a redefinition of what rationality is, at least as far as the direction of complex mechanisms is concerned.

The new hope for rationality in cybernetics seems obvious enough. The subject matter of cybernetics—the control of dynamic processes and the prevention of increasing disorder within them—was exactly the concern of Soviet administrators. Perhaps through the new science of cybernetics, they thought, genuine control of the immensely complex Soviet economy and government could be achieved. Whether that hope was justified and whether cybernetics is a true science are, of course, still unanswered questions.

The second result of cybernetics—the redefinition of rationality in controlling complex mechanisms—arose from the very nature of cybernetics. It is necessary, therefore, to spend a little time in defining the subject.

The Science of Rational Control

The term "cybernetics" is often improperly understood as being synonymous with automation. It brings to mind discussions of unemployment and impressive statistics about the number of operations a computer can perform in one hundredth of a second. In its original sense, however, cybernetics meant something quite different. The founders of cybernetics—Norbert Wiener, W. Ross Ashby, Arturo Rosenblueth, Claude Shannon, and John von Neumann—believed they were advancing a generalized theory of control processes. To them, a "control process" was the means by which order is maintained in any environment —organic or inorganic. In terms of this view of cybernetics, a computer by itself is not a cybernetic device. It can become a part of a cybernetic system when it is integrated with the other components of that system in accordance with a control theory.

The aspiring scientific discipline of cybernetics did not base itself upon the technological innovations which permit the construction of modern computers. Instead, it rested on the concept of entropy, taken from thermodynamics and broadened to mean the amount of disorder in any dynamic system. According to this approach, all complex organisms are constantly threatened by an increase in disorder, with the endpoint complete chaos. However, certain organisms are arranged in such a sophisticated and efficacious manner that through a dynamic process they can resist, at least temporarily, the tendency to disorder. Cybernetics studies the common features of these organisms and particularly their use of information to counter disorder. The more enthusiastic supporters of cybernetics view human society, which also obviously places a premium on order, as a particular type of cybernetic organism. In sum, cybernetics is the science of control and communication directed toward fending off increasing entropy, or disorder.

Cybernetics is a materialistic doctrine. It also postulates that the control features of all complex processes can be reduced to certain general principles. Yet its mode of operation differs distinctly from the science of the eighteenth and nineteenth centuries out of which the scientific optimism of Marxism arose. In the terms of the Enlightenment, rationality came through the knowledge of laws which would permit the prediction of the future. Such rationality was perhaps best symbolized by the celestial mechanics of Laplace. Control of a process, according to this early view, was based on knowledge of all physical laws and variables needed to predict future states of the process, and on the ability to change the magnitudes of the variables. Even the indeterministic nature of modern physical theory did not destroy the belief that rationality is essentially a theoretical rather than an empirical approach. In economics this concept of rationality led to the belief that if a centralized economy were not running smoothly, the difficulty must be inadequate knowledge at the center of local conditions and of the necessary economic laws for the changing of these conditions.

Cybernetics—which is based on analogies among all complex self-perpetuating processes, with living organisms the ultimate examples of success in self-perpetuation—

does not emphasize exact prediction of future states or conditions. Nor does it call for strict centralized control. The executive or command organs in all truly sophisticated cybernetic mechanisms are arranged in hierarchies of authority, with clearly autonomous areas. Furthermore, rather than trying to predict indefinitely the results of its executive actions, a cybernetic system makes constant empirical checks of these results through feedback, and it adjusts its commands on this basis. As Norbert Wiener said, cybernetics derives from control on the basis of actual performance rather than expected performance. Cybernetics thus places a premium on combining two things: local control based upon empirical evidence, and overriding centralized purposes.

It is a mistake to believe that cybernetics makes it possible to control the most complex processes by collecting in a central location enormous amounts of information. Indeed, cybernetics holds that barriers to information matter as much for control of processes as do free-flowing avenues of information. The best example of this paradox can be found in the human body, in many ways the paragon of a cybernetic mechanism. If we were conscious of everything that goes on in our stomach, or even just that information which some part of the body must be aware of in order for proper digestion to take place, we would be very neurotic indeed. Yet the human body represents the greatest victory of control over a complex process to which cybernetics can point; the features of its organization are basic to an understanding of cybernetic systems.

The Rebirth of Hope

The lesson of cybernetics for the Soviet Union, and especially for its economy, seemed clear. If Moscow knew everything occurring in its factories in Omsk, it would

also be "neurotic," as indeed it was when it attempted
to do so. The point is not that cybernetics was the cause
of the decentralization of the Soviet economy in the late
1950s. That reform was implemented prior to the recognition of cybernetics, and flowed from doubts about past
Soviet planning. Cybernetics restored faith in over-all
planning by presenting clear analogies for the combination of decentralization with overriding central purposes.

Cybernetics revitalized, at least temporarily, the leaders' confidence that the Soviet system could control the
economy rationally. This renewal came exactly at the
moment when the possibility had seemed to be irretriev-
ably vanishing. This rebirth of hope is the explanation
of the recent intoxication with cybernetics in the Soviet
Union.[1] Since 1958 several thousand articles, pamphlets,
and books on cybernetics have appeared in the Soviet
press, and the flow continues at an extremely rapid rate.[2]
In the more popular articles the full utilization of cybernetics has been equated with the advent of communism
and the fulfillment of the Revolution.[3] If the curious mix-
ture of ideology and politics in the Soviet Union can upon
occasion suppress certain sciences—as it did until recently
with genetics—it can also catapult others to unusual prominence.

One can find no other moment in Soviet history when
a particular development in science caught the imagina-
tion of Soviet writers to the degree which cybernetics has.
Perhaps the closest parallel occurred in the 'twenties, when
GOELRO, the State Commission for Electrification, was
made the subject of poetry. At that time, too, the industrial time-and-motion studies of Frederick Winslow Taylor
were applied widely and somewhat indiscriminately, and
the general enthusiasm for industrialization expressed itself on occasion in such unusual forms as concerts for the
workers in which the instruments were factory whistles.

But even the 1920s will not serve as a parallel. For cybernetics is being held out by its most ardent advocates as a far more universal approach than any of the diverse theories of the 'twenties.

It is quite common in the Soviet Union to find articles on the application of cybernetics in such surprising fields as musicology and the fisheries industry, although fre quently such expositions involve distortions of the meaning of the term cybernetics.[4] The normally stolid and reserved academicians of the Academy of Sciences have been among the most exuberant disciples of the new field. The Communist Party itself, since 1961, has endorsed cybernetics as one of the major tools for the creation of a communist society.[5]

The combination of centralized purposes with decentralized organization in a national economy obviously contains contradictory strains. A number of Western commentators have observed that the degree of success which the Soviet Union obtains in the one direction will be accompanied by the corresponding degree of failure in the other. There is no doubt much truth in this observation. Soviet administrators have not been alone, however, in being attracted by the new possibilities of combining what appear to be opposing principles. The United States is the country that supplies Soviet administrators with models for the organizational forms of industry, just as it did before World War II. The rhetoric in the following description in an American text of the organization of the Sylvania corporation could be transferred to a Soviet publication with no difficulty:

> In a revolutionary hookup Sylvania has 12,000 miles of communication network connecting 51 cities, to produce what spokesmen for that company call a step in "administrative automation." Their network of operation aims ultimately to include payroll, supply maintenance, auditing,

statistical services and the byproducts thereof, in their various plants, sales offices, and warehouses. This form of integration secures many of the advantages of centralized control in decentralized locations, a feat which previously seemed tantamount to having one's cake and eating it too.[6]

If cybernetics provided a stimulus for the hopes of Soviet planners, it did so in a rather ironic fashion. Traditionally, Soviet leaders have maintained that the key to their progress was the socialist organization of society; they now frequently come close to saying that science will *permit* the organization of such a society. Soviet economic planners during the 1930s spoke of how the communist state would provide for the welfare of science, a part of the cultural superstructure; they now speak of how science, which according to the Party is becoming a direct productive force and therefore a part of the economic substructure, will provide for the welfare of the communist state.

In other words, while in the 1930s we could speak of the Bolshevization of science, we can now speak of the scientization of Bolshevism. To the multitude of problems faced by the leaders of the Soviet Union—the mechanics of planning an intricate economy, the need for producing ever greater quantities of agricultural and industrial goods, the subordination of a land with climatic and other natural disabilities—science seems to provide the only possible solutions. The products of physical and social science research present the last opportunity for the Soviet state to be as rational, or its fruits as bountiful, as the early partisans of the centrally-planned state hoped.

Ideological Discussions

Cybernetics coincides with the materialism and optimism of Marxism, but it also raises a number of serious philosophical and sociological problems. Cybernetics is closely connected to a number of fields—such as psychology, econometrics, pedagogical theory, logic, physiology, and biology—which were subjected to ideological restrictions in Stalin's last years. In the early 1950s Soviet ideologists were definitely hostile to cybernetics, although the total number of articles opposing the field unequivocally seems to have been no more than three or four.[7] One author, writing in *Literaturnaia gazeta* in 1952, termed cybernetics a "science of obscurantists" and ridiculed the "mechanistic" view that computers can think or duplicate other functions of organic life. The militant critic accused the leaders of capitalist society of promoting the development of cybernetic mechanisms which would perform their society's unpleasant tasks for them: the striking and troublesome proletariat would be replaced by automatic machinery, bomber pilots who object to bombing helpless civilians would be replaced by "unthinking metallic monsters."[8]

Cybernetics seems to have found its first defender in the Soviet Union in Ernst Kolman, the Czech philosopher and mathematician. Professor Kolman, a long-term resident of Moscow, played a revealing role in disputes over the philosophy of science. Among Czech scientists he is generally known as a rigid ideologue but, perhaps not so paradoxically, in the Soviet Union he has often taken the more liberal side in various controversies. As long ago as 1938 he was praised by the eminent Soviet physicist V. A. Fock for his tolerant view of certain philosophic implications of relativity theory. In a recent issue of *Voprosy filosofii* Kolman pleaded that Soviet scientists be given permanent freedom to consider scientific theories which contradict common-sense notions.[9] In 1954, Kolman gave a lecture on cybernetics to the Academy of Social Sciences of the Central Committee of the CPSU, which was very much in this tolerant tradition.[10] Kolman empha-

sized his belief that cybernetics was causing a technological revolution which the Soviet Union had so far largely
ignored. This revolution could be compared in significance, he said, to the implementation of the decimal
numeral system or the invention of printing. Only later
would the full irony of Kolman's assuming the role of
champion of cybernetics emerge; in later years he was
exceeded in his enthusiasm for the field by many Soviet
scholars, and in a number of recent articles he has ap-
pealed for restraint in evaluating the applicability and
potentiality of cybernetics.[11]

In the period from 1954 to 1958, Soviet scholars de-
bated the legitimacy of cybernetics as a field of study. A
number of them noted, quite correctly, that there were
no accepted definitions of "cybernetics" or "information"
even in the West; much of the discussion revolved around
fruitless attempts to agree on terms. A clear trend in the
discussion soon emerged, however. Leading Soviet scientists increasingly shifted from criticism of the new field
to its defense. The prominent mathematician A. N. Kolmogorov illustrated this trend with his initial refusal to
recognize cybernetics followed by his declaration at a
meeting of the Moscow Mathematical Society in April
1957 that he had previously not understood the genuine
meaning of cybernetics and now supported it.[12]

The movement toward cybernetics grew gradually but
continually until it began to assume the dimensions of a
landslide. In April 1958 the Academy of Sciences of the
U.S.S.R. announced the creation of the Scientific Council
on Cybernetics, headed by Academician A. I. Berg, which
included such diverse members as physicists, biologists,
lawyers, mathematicians, economists, and linguists. The
Academy's Institute of Automation and Telemechanics
began directing most of its research toward what it de-

scribed as cybernetic applications. Computer centers were established in some of the Academy's branches.

In the Soviet press, students were urged to major in cybernetics. Soviet science fiction was filled with utopian descriptions of "cybernetic brain-modeling" and "cybernetic boarding schools." As if to illustrate that this future was not far away, the Academy of Pedagogical Sciences of the Russian Republic established an experimental boarding school to prepare children for careers in cybernetic programing.

In the discussions in the Soviet Union, both before and after the official endorsement of cybernetics in 1958, the ideologically most controversial question concerned the realm of applicability of cybernetics. The spectrum of the debate ranged from those who believed cybernetics to be no more than a loose word for process engineering to those who saw it as a new science providing the key to literally every form of the existence of matter. The question of the realm of applicability contained many aspects. The debate over whether machines can "think" and the nature of cybernetic information were parts of the general issue of the applicability of cybernetics.

The Soviet interpreters of cybernetics who granted it the widest range of applicability referred to the writings of the Western scientists Wiener and Ashby, who often spoke of the homeostatic properties of society.[13] They also cited such Western political scientists as Karl Deutsch, who attempted to construct cybernetic models of political behavior. The ideologically more conservative and orthodox in the Soviet Union thus believed that cybernetics was a possible rival to Marxism, which also advances a general scheme to explain both natural and social sciences.[14] As one alarmed Soviet author commented, "The attempts to convert cybernetics into some kind of universal philosophical science are completely groundless. Marxists will reject them out of hand."[15]

A similar problem of the applicability of cybernetics concerned the question: "Can a machine think?" More exactly phrased, this question is usually posed as: "Do computers perform tasks which are merely analogous to thinking, or do they possess structural identity with it?" Some prominent Soviet scholars who first supported cybernetics—when it was still very controversial to do so—have since moved into the conservative wing on the question of the ability of computers to think. Academician Berg, for example, unhesitatingly commented, "Do electronic machines 'think'? I am sure that they do not. Machines do not think and they never will think."[16]

The question of whether or not computers can in principle duplicate mental functions is still a very controversial one in the Soviet Union, with adherents on both sides. Yet it seems unlikely that an affirmative answer to this question will receive ideological support. Systematic dialectical materialism may not specifically deny the possibility of thinking machines, but the anthropomorphic nature of historical materialism is a genuine obstacle to such an admission.[17] A number of Soviet authors have recently maintained that the essential difference between man and machine is not technological, but social. As Kolman commented, "Those who maintain that man is a machine and that cybernetic devices think, feel, have a will, forget one 'trifle'—the historical approach. Machines are a product of the social, or work, activity of man."[18] A like view, even more forcefully expressed, was presented by N. P. Antonov and A. N. Kochergin: "It is necessary to emphasize that man works and not the machine. One can say that the machine functions, but not that it labors. The machine cannot become the subject of laboring activity because it does not and cannot be possessed of the necessity to work, and it has no social requirements which it must labor to satisfy. This is the main and principal difference between machine and man."[19]

A question even more important than the ability of machines to duplicate man's functions is that of the moral responsibility of man for the actions of his machines. On the whole, Western cyberneticians are, at least publicly, more fearful than Soviet scientists of the possible results of their employment of computers. In 1960, when Norbert Wiener visited the editorial offices of the leading Soviet journal in philosophy, *Voprosy filosofii,* he commented:

> If we create a machine . . . which is so "intelligent" that in some degree it surpasses man, we cannot make it altogether "obedient." Control over such machines may be very incomplete. . . . They might even become dangerous, for it would be an illusion to assume that the danger is eliminated simply because we press the button. Human beings, of course, can press the button and stop the machines. But to the extent that we do not control all the processes which occur in the machine it is quite possible that we will not know when the button should be pressed. Thus, the programing of "thinking" machines presents us with a moral problem.[20]

Wiener's uneasiness has been expressed in different terms by other Western cyberneticians who have spoken of the possibility of a dictator controlling society through the use of cybernetic machines, while still others have referred to the computer as the demon which turns upon its master.

All these pessimistic Western views are rejected by Soviet writers. Like the ideologists, Soviet scientists are, with very few exceptions, eternally optimistic in their public statements about science. If any of them have, in Oppenheimer's phrase, "come to know sin" as a result of their research, they keep this encounter to themselves. Indeed, several Soviet scholars have expressed the view that the essential difference between man and machine is the fact that man sets his own goals while the machine strives only toward those for which it has been programed. If society places a premium on worthwhile goals, say the Soviet authors, the machines of that society will be assigned similarly meritorious functions. These writers suggest that Western cyberneticians betray a lack of confidence in their own societies when they doubt the roles their computers will be asked to play.

The most extreme supporters of cybernetics in the Soviet Union have been scientists and engineers such as Academicians Kolmogorov and Sobolev. Kolmogorov, who for many years resisted the unrestrained claims of cybernetics, has become quite outspoken in his belief that computers can in theory reproduce human thinking. He has published articles with titles such as "Only an Automaton? No, A Thinking Creature!" Sobolev has taken an even stronger pro-cybernetics stand. In a debate with a reluctant philologist, Sobolev unhesitatingly called man a cybernetic machine. He spoke of the possibility of man's creating other machines which would be alive, capable of emotion, and probably superior to man.[21]

In 1961 and 1962 it appeared that a solution to the problem of competition between dialectical materialism and cybernetics as schemes with universal pretensions had been found. This "two-plane" solution emphasized a division of labor. Cybernetics was defined as the science of control and communication in complex mechanisms. Marxism, on the other hand, was the science of the broadest laws of nature, society, and thought. According to this line of reasoning, Marxism was so much more general an intellectual system than cybernetics that the two did not conflict.[22]

The two-plane approach of several years ago did not permanently solve the problem of competition between cybernetics and dialectical materialism. In recent articles in Soviet philosophical journals, several authors postulated that cybernetics analysis may be applied to literally all phenomena. They link "information" to the Leninist concept of "reflection," supposedly inherent in all matter. Most of all, the new aggressiveness of the cyberneticians goes back to technical definitions of information. [23]

The enthusiastic proponents of cybernetics have maintained that the evolution of matter, from the simplest atom to the most complex of all material forms, man, may be seen as a process of the accumulation of information. Thus, these authors tied together cosmogonical, geological, and organic evolution in one process of the tendency of matter, at least in certain loci, to increase its informational content. The engineer E. A. Sedov and the mathematician A. D. Ursul advanced the proposition that through such an understanding of information the competence of thermodynamics would be extended over all matter, including life. These authors consider an increase in orderliness resulting from a cybernetic process of self-preservation to be a natural evolutionary tendency in the material world "beginning with the origin of the galaxies and planetary systems" and ending with organic structures.[24]

Articles such as Sedov's and Ursul's do not agree with the Soviet definition of cybernetics which allowed its reconciliation with dialectics. That two-plane definition limited cybernetics to control processes, while dialectics was granted a universal domain. Now, however, these and other authors are calling for a recognition of the validity of a cybernetic approach to all phenomena, and the uniqueness of dialectics again seems threatened, at least to the more orthodox philosophers. A literal rephrasing of the three basic laws of the dialectic in cybernetic terms was attempted by the author of an unpublished doctoral dissertation at Moscow University.[25] The uncertainty at the present time is revealed by the fact that a recent issue of *Voprosy filosofii* contained directly contradictory statements on the universality of cybernetics.[26]

Cybernetics and Marxism: Conflict or Concurrence?

It is very tempting and convenient to see the cybernetics discussion as one more episode in the sequence of conflicts between dialectical materialism and science in the Soviet Union. As in the past, one might argue, the final result has been a blow to the status and prestige of Marxism-Leninism. Without any question the cybernetics controversy illustrates clearly the general tendency in the Soviet Union since Stalin's death to shift from ideologists to scientists for reliance in interpreting science. Yet cybernetics has been a double-edged sword. Cybernetics has caused a great deal of nervousness for the ideologists, especially those held over from the Stalin era. Yet it has also given a remarkable boost to the beliefs in rationality, scientific optimism, and materialism—primary characteristics of the official Soviet mentality. In so doing it has provided the new generation of Soviet philosophers, some of whom possess considerable scientific training, with a topic of investigation containing both philosophic and scientific stimulation. Cybernetics has literally been both a threat and a boon to Soviet ideology, and my opinion is that in the final analysis it has provided more assistance to Soviet hopes than it has contributed to the erosion of the Soviet world view.

To explain this position it is necessary to make a few comments about the nature of Soviet ideology and particularly of dialectical materialism. What relevance does dialectical materialism as the general Marxist view of science have to any specific scientific field like cybernetics? We often assume a direct connection. We are quite accustomed to saying, for example, that the controversy over Lysenko was a genuine blow to the strength of dialectical materialism. On close examination it becomes clear that the important issues in the Lysenko controversy have no direct connection to dialectical materialism. There is nothing in the formal framework of dialectical materialism about the inheritance of acquired characteristics, intraspecific competition, or the phasic development of plants. By the same token, there is nothing in dialectical materialism which inherently contradicts or affirms cybernetics.

Systematic dialectical materialism consists of a set of very abstract statements about what is termed objective reality. It contains a theory of knowledge, the copy-theory, and an ontology of change or movement, the laws of the dialectic. This set of statements is flexible almost to the point of evanescence. It has been used both for and against certain interpretations of quantum mechanics, relativity theory, and genetics. It may contain one or two principles which could not be surrendered without destroying the system, such as the belief in objective reality. But these principles are so general that they are immune from refutation, given any skill in interpretation and rationalization. In recent years such skill has been highly developed in the Soviet Union.

Hence, when Western observers say that the Lysenko affair discredited Marxism-Leninism, they usually do not mean that dialectical materialism was found incorrect in its basic principles. Rather, they have in mind that the prestige of the Communist Party, which committed itself to a pseudo-science, was severely damaged. Similarly, what I mean when I suggest that cybernetics has given, in the balance, a boost to Soviet ideology, at least in the minds of its adherents, is not that it verifies dialectical materialism. What matters here is that the Communist Party's half-century-old claim that human society can be given a rational direction has been renewed. The fact of this renewal is as important as the validity or lack of validity of cybernetics as a means of reaching that goal.

The Destiny of Cybernetics

If Lysenkoism, which included the idolization of a certain kind of "science"—Michurinism—resulted in a blow to Party ideology when all the poverty of that position was revealed, would not a loss of faith in the power of cybernetics also result in an erosion of ideology? As the likelihood of a drop of Soviet enthusiasm in cybernetics is, in my opinion, very great, this difficult question is quite pertinent. Most people who have encountered cybernetics as a conceptual scheme have gone through a cycle of intoxication, or at least stimulation, followed by a period of disillusionment, or at least retracted ambitions.

For all the persuasiveness of cybernetics upon first contact, it is a very incomplete science.[27] Cybernetics may dissolve into less dramatic subareas of information theory and computer technology. As a French specialist in cybernetics observed, "As an adjective, 'cybernetic' threatens to go the way of 'atomic' and 'electronic' in becoming just another label for the spectacular."[28] Certain scientists find the use of the term embarrassing. Furthermore, it is now clear that there were genuine defects in the writings of several of the founders of cybernetics who, in their enthusiasm, often confused certain technical terms, such as "quantity of information" and "value of information."[29] And, finally, cybernetics proceeds on the basis of analogical reasoning, which by itself leads not to logi-

cal or scientific proofs but instead to inferences which
may or may not be significant and fruitful.

The strength of such reasoning depends upon the
similarities that can be identified between the two entities
being compared. Soon after the development of the methodology of cybernetics, the comparison of the human
body as a control system to an economic system, a city
government, or an automatic pilot seemed to result in
the identification of truly striking similarities. The longer
one dwells upon such analogies, however, the more clearly
emerge the very genuine differences which exist between
the entities being compared.

There are cyberneticians, such as Stafford Beer, who
maintain that cybernetics goes far beyond analogy. They
argue that, if one abstracts the control structures of two
dissimilar organisms, the relationship between these structures may be one of identity rather than of analogy.[30]
The control structure of a complex industry and that of a
living organism may be identical, according to this view,
in the same way that the geometrical form of an apple
and that of an orange may be identical circles. This approach may be true on an abstract level, but it has not
resulted in as many discoveries of fruitful similarities and
avenues of research, beyond those originally identified,
as early proponents of cybernetics hoped.

In the history of science we are familiar with an exuberant seizing upon the latest conceptions of science as
models for behavior in areas rather distant from their
points of origin. It is not accidental that the development
of Newtonian physics was followed by the application of
such concepts in a universalist fashion, with the human
body described as a mechanical assemblage and the relations of European states being analyzed by the use of
such terms as "balance of power" and "fulcrums" of diplo-
matic pressure. Such examples could be multiplied easily.

In the 1930s the concepts of "complementarity" and "indeterminism" arising from the study of subatomic particles were applied by certain physicists and philosophers in attempts to explain cultural relativism and freedom. With the development of cybernetics, our vocabulary is enriched with models of political behavior containing "feedback" and with models in which the achievement of "homeostasis" is the desired goal.

Through the history of science, each of these various models has been of scientific importance. Each has also been of heuristic value in fields other than the science in which they were originally developed. Some, of course, have been more valuable than others. But all have been followed by periods of greater appreciation of the failures and inadequacies of such models as well as of their successes. Every model of behavior emphasizes certain attributes of an organism at the expense of others. The greatest heuristic value of such models usually occurs at the first moment of genuine understanding.

In cybernetics, therefore, the absence of dramatic theoretical breakthroughs is apt to lessen the persuasiveness of its conceptual scheme as an explanation of all dynamic processes. In the United States, where computers are applied very widely and where their sociological and economic consequences are still topics of vigorous debate, the decline in interest in cybernetics as a conceptual scheme is clearly evident. The post-cybernetic epoch involves not a renunciation of cybernetics but only a more sober appraisal of its potentialities. The original zeal might be renewed by future developments in theory, but one obviously cannot foretell such events.

In the Soviet Union cybernetics has achieved remarkable stature. If we assume for the purpose of discussion that theoretical breakthroughs do not occur in the near future, what would be the effect of the resultant decline of cybernetics on ideology in the Soviet Union? In the first place, the result would not be anything comparable to the loss of prestige that the ideology faced with the discrediting of Michurinist biology. The differences between the newest idol of Soviet ideology and Michurinist biology are clear. Lysenkoism had no theoretical base, while cybernetics has a broad theoretical base which might be judged as incomplete but not as incorrect. Lysenkoism was a counter-science, opposed to Mendelian genetics, and it centered in one country; cybernetics is not opposed to any existing discipline and it is a subject of research in many countries. The greatest ideological boon in cybernetics, so far as the Soviet Union is concerned, is its optimism and rationalism. Lysenkoism was optimistic but thoroughly irrational. The failure of Lysenkoism was the failure of pseudo-science. The failure of cybernetics would mean a delay, which many would say is temporary, in man's ability to control the next order of complex processes.

Whatever the destiny of cybernetics as a science, I would foresee in the Soviet Union a lasting residue of renewed belief in man's ability to control intricate processes based on the new wave of achievements in this area. Thus, the scientific optimism to which Marxism is an heir and to which the Soviet ideologists have constantly striven to tie their particular social system has received a new impetus. Notice that I have not concluded that ideology as a whole has increased in vitality in the Soviet Union. I merely conclude that the result of one particular phenomenon, the rise of cybernetics, has been a boost to the ideology.

Since Stalin, in my opinion the Soviet Union has witnessed some trends which erode and others which sustain ideological commitment. In the area of literature, for example, the prevailing trend unquestionably runs counter to the maintenance of the ideology as defined by the Party. It may well be that in the long run the renewed scientific optimism which I have stressed cannot alone sustain the ideology, for the more closely ideology is tied to science, the less the ideology can be separately identified. But in the foreseeable future the faith in both science and socialism, coupled with the political power of the Soviet Union and its ruling Communist Party, will be a factor sustaining rather than eroding the ideological commitment.

ECONOMICS

HERBERT S. LEVINE

The period since the late 1950s has seen a veritable explosion of Soviet interest in the use of mathematical and computer techniques in economics. Hundreds of articles and books have been written, research institutes and government committees have been established, numerous conferences have been held, courses on mathematical methods have been introduced at all leading universities. And Lenin Prizes, of great symbolic importance, have been awarded to three outstanding mathematical economists.[1]

New scientific approaches, of which mathematical

techniques are one aspect, have been accepted more fully in economics than in other social sciences in the Soviet Union. Most likely, this is due largely to the higher degree of quantification found normally in economics. Not that Soviet economics employs an exceptionally high level of quantification; quite the reverse, a major complaint of Soviet mathematical economists today is that many concepts that should be expressed in terms of quantities are still expressed in terms of qualities and thus are not amenable to mathematical treatment.[2] But essentially the technical relationships in the economy are measured quantitatively and planned quantitatively, and the results of economic activity are measured in quantitative terms. Furthermore, in recent years the slowing down in the rate of economic growth, clearly evidenced by quantitative measures of economic performance, has led the regime to seek new approaches in economic planning and control. In these approaches mathematical methods play a significant role.

In this chapter I explore the impact that the mathematical revolution in Soviet economics has had on the planning and operation of the Soviet economy. I also consider the impact that it has had, or may yet have, on the role played by ideology in the economy and in the science of economics.

In the first section, I give a brief picture of the present Soviet system of constructing and implementing an overall economic plan. This picture serves as a background for discussing, in the second section, the possible ways in which the system could use mathematical techniques in economic planning and control. The third section indicates in a general rather than detailed way what uses Soviet economists have made so far, in actual practice, of these techniques; the impression is one of slower progress than was expected. The fourth section suggests some reasons for this divergence between hopes and reality. In the concluding section, besides drawing some general conclusions, I try to link this chapter with the preceding ones in this volume by commenting on the compatibility, in Soviet economics, of the new scientific approaches and ideology.

Constructing and Implementing the Plan

Soviet economists construct a number of different types of economic plans,[3] the most important of which are: a long-term plan, for five years, and a short-term plan, for one year. To some extent, these plans are constructed in similar ways. But the annual plan is the operational plan in the Soviet economy, in the sense that the numbers it contains form the basis of commands to economic units. Hence we will focus on the methods of constructing this short-term plan. It is constructed by means of a flow-counterflow process. This means that instructions flow down the planning and administrative hierarchy, and information flows up.

The political leaders communicate their preferences, which are dominant in the system, to the planners. On the basis of these communications and of at least part of the multitudinous statistical data collected within the economy, the planners then prepare a tentative, highly aggregative set of output targets and input limits. These preliminary instructions are sent to the ministries. The ministries in turn subdivide them among subordinate organizations; ultimately they reach the individual firms. The firms then initiate the counterflow of information. On the basis of the tentative instructions it has received, each firm is supposed to prepare a more detailed list of its production possibilities and needs. The list shows two things: the levels of output of its various products that it thinks it can attain, and the inputs it feels it needs to attain them. These lists are sent to superior organizations, argued out, and consolidated at the ministerial level. They are then sent to Gosplan (State Planning Commission) where, after more arguing and consolidating, economists work out the internal consistency of the plan—that is, the necessary relationships between inputs and outputs. The Gosplan draft of the plan is then sent to the Council of Ministers for alteration and confirmation.[4] The confirmed plan becomes a legally binding document, the Annual Plan for the Development of the National Economy of the U.S.S.R. On the basis of it, more detailed production plans are made and relations between firms are established.

For the purposes of this chapter, three features of the process of constructing the plan warrant special note. The first concerns the level of detail in the plan. The preliminary plan has customarily covered the *planned* part of the economy at a level of aggregation of roughly 200–300 product designations. The confirmed plan has usually employed a level of aggregation in the range of 800 to 2,000 product designations for the output and distribution of the most important economic items. The detailed plans that are constructed on the basis of the over-all confirmed "state" plan have used, in recent years, a level of aggregation of up to 18,000 product designations. The difficulties of centrally constructing a 2,000-order plan are considerable; the difficulties of coordinating plans for 18,000 products are monumental. Yet the Soviet economy in actual operation uses product designations for different qualities, sizes, and shapes that number in the millions. At best, therefore, any central plan can be balanced only in terms of larger groups of products, and inconsistencies can easily develop when the detail necessary for operational purpose is added.

A second noteworthy aspect of constructing the plan is the role played by the input norm. An input norm is an upper limit on the amount of an input that may be used in the production of a unit of any output. Input norms are established for the important inputs and outputs of a firm at various levels of the planning hierarchy and then are used by the firm to determine its input needs for its proposed list of output targets. In a certain sense, they take the place of prices in the firm's decisions on choice of production techniques. But they cannot effectively perform the functions of prices because they are based primarily on engineering factors, with little consideration given to economic factors—that is, to the relative scarcity of the various inputs.

Finally, the method used by Gosplan to work out the internal consistency of the plan contains inherent difficulties. For each centrally planned product, Gosplan constructs a balance sheet called a material balance. On one side, the planned sources (production, imports, inventories, etc.) of the product are listed, and on the other side, the planned uses (production, construction, reserves, etc.). As information comes in from the ministries and is accepted by Gosplan, it is communicated to the specialists on the relevant material balances, who record the data in the appropriate places on the balance sheets. When all the information is in, each side is totaled and compared with the other. It is only by the wildest chance that the two sides will be equal at this stage.

Usually the planned demands (uses) are greater than the planned supplies (sources). The problem then is to bring all the material balances (800 to 2,000) into simultaneous balance. This is very difficult for if the output level of a product is increased, where the planned demand for it exceeds the planned supply, then the output levels of all the inputs into this product have to be increased, and then the output levels of all the inputs into the inputs, and so on. The extent of the indirect consequences of such an initial change makes that way of solving the problem too difficult to be handled by Gosplan, which employs nothing more powerful than desk calculators for its balancing job. For some of the most important products, Gosplan does trace through the secondary effects for a level or two. But, in general, Gosplan relies on balancing methods which avoid secondary effects. What they do, mainly, is reduce the input norms. Producers of products in short supply are told to increase output but without using any more inputs; likewise, users of such deficit products are told to maintain their levels of output but with less of an allocation of the deficit input. This tightening of input norms has the advantage of putting pressure on producers for high levels of output. But when it is overdone it results in unrealistic and unrealizable plans.

Without evaluating in detail the effectiveness of Soviet methods of constructing their economic plans, it can be said that Western analysts—and in recent times Soviet analysts, too—agree that a plan constructed in this way: (1) is usually late, so that in the beginning of the year, the economy operates without a plan; (2) is, for the reasons noted immediately above, almost inevitably not internally consistent; and (3) is non-optimal, because the relative scarcity of potentially substitutable inputs not having been determined, the chance to increase the output level of some products without decreasing the output levels of others has been missed.

The second part of the planning process involves implementing the plan. Much of what is meant by the term "command economy" applies to plan implementation.

Soviet planners communicate the targets in the plan to the executors of the plan (the ministries and, below them, the producing units) in terms of direct commands: "You will produce at least a million tons of steel." To foster the high rates of economic growth demanded by the political leaders, the management of the firm is given incentives to fulfill and overfulfill the plan, in the form of penalties for failure (primarily loss of managerial status and its perquisites), and rewards for success (monetary bonuses). These bonuses used to be related mostly to the fulfillment and overfulfillment of the physical output target, but, under the new system announced at the Party Plenum in September 1965, they are to be related to the volume of sales. More attention is also to be paid to the profit earned by the firm.

Control over implementing the plan is exercised by organizations superior to the firm inside and outside of the ministry, including local organs of the Party. This control takes many forms: the redirection of inputs, the sending out of reserve inputs that are kept by the center, and the repeated changing of plans during the course of plan fulfillment.

Much of the harsh criticism that Soviet officials and specialists directed, before September 1965, against their system of economic planning and control relates more directly to the problems of plan fulfillment than of plan construction. The same applies to much of the economic reform that has been suggested, including the oft-cited proposals of Professor Yevsei Liberman. The major complaints have been that the commands given to firms are too numerous, too detailed, and often contradictory, and that the incentive mechanism used before September 1965 gave rise to many actions on the part of the firm which ran counter to the aims of the economy.

The Uses of Mathematics: Theory

Given the above-described planning methods and the problems they produce, we may now ask exactly what could the Soviet economy gain from the use of new mathematical and computer techniques?

Once Soviet economists themselves began to discuss this question, a tone of urgency quickly developed. While Soviet planning methods had not really changed much from the beginning of the 1930s, the scope and complexity of the economy had changed drastically. One Soviet economist complained that although everyone knows that a hydroelectric power station cannot be built with a shovel, not everyone realizes that "such a complex economy as ours cannot be directed with the aid of abacus, paper, and pencil."[5] The most famous description of the problem is the oft-quoted statement (outside Russia *and* within) of the prominent mathematician-cyberneticist, V. M. Glushkov, to the effect that planning work increases by an amount equal to at least the square of production, and therefore, if nothing is done to modernize the planning system, by 1980 planning work will occupy the entire adult population of the Soviet Union.[6]

One of the first areas in which Soviet economists saw a possible use of mathematical methods and especially of electronic computers was that of information systems. An information system involves the collection, transmission, processing, storage, and use of data. And such systems are vital in the Soviet economy for the constructing of plans and control of their fulfillment. Although it has long been known that Soviet economic information systems were cumbersome and inefficient, the horror tales Soviet economists have been telling in recent times are, nonetheless, shocking.[7]

Soviet government agencies collect mountains of information. Much of it seems to be inaccurate, irrelevant, and duplicative. In fact, since different ministries and other organizations use different ways of classifying information, they collect noncomparable mountains of information. They then process the information so poorly and store it so inconveniently that only a fraction of it is over used.[8] Professor Richard Judy, in a study of Soviet economic information systems, characterized the situation:

> From this survey of existing Soviet economic information systems, it seems obvious that a major information crisis faces Soviet economic management. Each of several parallel information systems imposes its own burden of reporting upon the enterprise. . . . There is little standardization and collaboration among the parallel systems. Voluminous files are kept separately in each system in spite of great commonality of the data items contained in them. Proper file maintenance is difficult because of poor file organization, inadequately designed inflow of updating information, and manual methods of processing. For the same reasons, information in the files can be retrieved for management use only with difficulty and after prolonged delay. Pertinent information is frequently not available to decision-makers even though managerial personnel spend most of their time in routine data-processing activities. Lack of information and the delays in its availability diminish the system's ability to respond to change. In short, information is incomplete, insufficiently accurate, largely irrelevant, and slow [and] expensive. . . . Like the talking donkey, the wonder is not that the system works badly but that it works at all.[9]

It is hardly surprising that Soviet economists, faced by such imperfections in the current system of data collection and handling, looked hopefully toward the use of computers, with their speed, greater reliability, and ability to produce data on demand, and that they considered one of the main virtues of mathematical methods to be that mathematics was the "language" of computers. Articles advocated a unified, multistage computer network, in which individual firms fed their information into local computers, which in turn fed sectoral and regional computers, until eventually all data were accumulated in central computers.[10] Such a network would make data available for planning, decision-making, and control at various levels of the economy, and of course would greatly speed up the flow of data.

Another use foreseen for computers was in the construction of input norms. As hitherto done, this was indeed a laborious and time-consuming job, because of the volume of data that went into their construction. For example, obtaining the annual-plan norms for the Ural Machine Building Factory required compilation of a document 17,000 pages long.[11] The job, Soviet economists said, could now be done on computers, freeing economists, statisticians, and accountants for more useful work and producing better, more up-to-date norms.

Some discussions of computerized data systems even looked toward a highly centralized cybernetic system of planning and control in which the computer would perform all plan elaboration and management decision-making. I shall refer later to cybernetics and to the options between centralization and decentralization.

A more likely and highly promising use for mathematics and computers was in the field of "input-output" methods. This method puts the technical interrelationships existing among products into a single scheme (matrix) and expresses them mathematically in such a way that the entire system (set of equations) can be solved on a computer. Starting with a set of direct input coefficients (norms), the input-output solution provides: a consistent set of total outputs (final plus intermediate outputs), a

set of final outputs (goods going to consumption, investment, and government, and exports minus imports) and
a set of what are called direct and indirect input coefficients (for example, the "direct" coal that goes into the
production of an automobile, and the "indirect" coal that
goes into the production of the steel that goes into the
production of the automobile, etc.).[12]

Early support for the use of input-output methods was
expressed in an important article by V. Belkin, and in the
publications by the great figure of the period, the man
who was the moving force behind the mathematical revolution, the late Academician V. Nemchinov.[13] Soon many
other writers were following their lead. The input-output
method appeared to be tailor-made for the problem of
giving the Soviet economic plan the internal consistency
it needed. The method would not require—at least not on
the surface—any change in approach. All that would be
necessary would be to substitute computers and input-
output matrices for the material balances, desk calculators,
and telephones of the Gosplan officials. The plan thus
constructed would be internally consistent. And although
the optimality of the plan would, as before, depend on
the considerations underlying the direct input norms used
in its construction, the input-output method could improve effectiveness by allowing the construction of some
variants of the plan. Once the computer was programed
and the basic data fed in, then there could quickly be
worked out a number of internally consistent variants of
the plan from which the political leaders could choose the
one they thought best. The substitution of the computer
for human labor would cut the time necessary for plan
construction at the balancing stage from about two months
to about two weeks. Thus the planners might be able to
get the plan to the firms before the beginning of the
plan year.

Soviet economists saw other potential advantages in

the use of the input-output method. With a set of direct and indirect input norms, the planners could rapidly calculate the total consequences of any change that the political leaders, at the confirmation stage or during the operation of the plan, wanted to make in the plan. Furthermore, if some things went awry during the plan year, the planners would be able to calculate the indirect effects of the changes to be made. Thus they could order changes that would maintain consistency in the plan rather than create contradictions. Finally, the input-output method was seen as playing a role in Soviet price formation: with the use of the direct and indirect input norms, a consistent set of prices based on labor values could be calculated. Indeed, it was the input-output method that appeared to be the first major mathematical tool that would be applied to Soviet planning practice.

Soviet economists also became interested in linear programing techniques. Whereas input-output is a method of achieving consistency in plans, linear programing reveals optimum solutions in plans. That is, input-output is a method that can be used to ensure a balance between planned outputs and planned inputs. But linear programing can be used to work out the maximum or minimum level of some measurable objective (say, level of output or cost) when there are possible alternative technologies and limited resources. The scope of linear programing problems can be as large, theoretically, as the entire economy or as small as an individual firm or a shop within a firm.

The first successful solution to a linear programing problem was worked out by the Soviet mathematician L. Kantorovich in 1939.[14] But it was not until the publication of his 1959 book[15] that Soviet discussions of the use of linear programing really began.[16] Most of the early discussions concerned the use of linear programing in limited problems where the objective was clear. Some problems involved the efficiency of processes in the firm, such as finding the way of routing blocks of metal through a number of machines which would maximize the output of the enterprise, or the way of cutting shapes from sheets of metal which would minimize waste of metal. Others involved efficiency of use of resources over a larger area, such as finding the way of transporting goods from a set of producing firms at various locations to a set of consuming firms in various locations which would minimize ton-kilometers of transportation.

More debatable was the advisability of using linear programing for broad economic problems where the objective was not clear and simple, but where, rather, major economic priorities were involved. Such problems, it was argued, were the domain of Party policy decisions. Certainly they should not be handled—as Kantorovich toward the end of the 1959 book seems to argue they should—by means of his "objectively-determined evaluators."

These "objectively-determined evaluators" were a key element in the attractiveness of the linear programing approach. Kantorovich was being tactful in not calling them prices. Yet in terms of Western economic theory that is exactly what they were, since they were measures of value in terms of opportunity cost. An "objectively-determined evaluator" measured the value of any input that was in limited (scarce) supply in terms of what was given up by not using the input elsewhere in the economy. Thus the "evaluators" gave to the Soviet economist for the first time a concept of price which was appropriate for rational economic decision-making.[17] In this lay their power. But in this also lay a certain ideological danger about which more will be said below.

The Uses of Mathematics: Practice

All these high hopes and great potentialities for mathematical methods might lead one to expect that at least some of the new techniques would have been fairly rapidly introduced into practice. In reality, very little—or as one prominent Soviet economist put it, "outrageously little"[18]—has been done so far to introduce mathematical methods and computer techniques into actual economic planning and management. Many models have been built both of a theoretical and of an applied character.[19] Much interesting work has been done, but for the most part it has been derivative in nature, being based on previous work done in the West.[20]

In addition to this model-building, there have been reports of considerable activity, at lower levels such as the firm and the region, in experiments with certain models and in the use of some computerized techniques of control and problem-solving of the operations research type. The problems attacked involve such matters as transportation, machine routing, plant location, livestock feeding, and crop distribution. But the new methods are confined to experimental cases and pilot models; Soviet economists frequently complain that even when the methods prove to be effective, they are not extended to other enterprises or to other organizations or regions.[21]

Soviet economists have spoken a lot about the design of a unified state network of computing centers, but little has been done about its construction. Recently a decree, "On improving the organization of work on the creation and use of computing equipment and automated control systems in the national economy," was issued by the Central Committee of the CPSU and the Council of Ministers of the U.S.S.R. This was followed by a sign that at least some organizational progress was being made: the task of setting up the network has been entrusted to a specific organization, the U.S.S.R. Ministry of the Radio Industry.[22]

What is perhaps most interesting of all is the fate that has so far befallen the use of the input-output method—that method where centralized, large-scale application of mathematical and computer techniques looked most promising, and which was especially applicable to Soviet planning methodology. Not only has there been no great achievement here, but a reappraisal on the part of Soviet economists of their early optimism seems to have taken place. More attention is now being devoted to the difficulties that lie in the way of the method's successful introduction into planning practice.[23]

Following construction of a number of small input-output pilot models, the first substantial work was completed in 1961. It consisted of an empirical input-output study (in physical units and in value terms) of the Soviet economy in 1959. Some parts of the value table were published, and scholars in the West reconstructed other parts, so that by and large a good picture of the table is available.[24] On the basis of the empirical 1959 table, Soviet planners constructed experimental *planning* tables in value terms and physical units, for 1962.[25] For the value table, only 500 direct input coefficients out of the 4,260 direct input coefficients in the table were corrected —that is, brought up to date—but these accounted for about 95 per cent of total material expenditures in the 1959 table. From these direct input coefficients, a set of direct and indirect input coefficients was derived. Final demands were taken from the official state plan for 1962 and broken into the sector groupings used in the table (this was reported to be a laborious job). The final demands were then multiplied by the direct and indirect input coefficients to obtain the planned levels of total output.

The planned levels of gross outputs (intermediate output plus final output) from the input-output model were compared with the planned gross outputs from the state plan. (Actually, their rates of growth since 1959 had to be compared, because the product compositions of groupings in the two plans were not the same.) It was found that, on the whole, gross output in the input-output model grew slightly more slowly than that in the official state plan. As both plans had the same levels of final output, this meant that the input-output plan saved resources which could then be used to increase final output. For many individual product groups the growth of gross output was the same; for some the input-output gross output growth was higher, and for others, lower. This showed excessive tension and therefore potential bottlenecks in the production of some products and reserves for increased final output in the production of other products.

The Soviet economists connected with this large study were apparently quite happy with the results. Even though the plan produced through the input-output approach was not very different from the one produced by means of the traditional material balances method, it did show that input-output could do the job at least as well as the old methods and could do it faster and (presumably) with less use of manpower. The economists involved looked forward to making input-output an integral part of the plan construction process. But it seems that the enthusiasm has faded somewhat since then. Planning tables were constructed for 1963, but again as an experiment and not too much was said of them.[26] The construction of planning tables for 1964-1965 was noted, but nothing has been seen of them.

In conjunction with the new five-year plan, a large input-output planning table (129 sectors) for the end year of the plan, 1970, was constructed by the same group that put together the 1962 value table. Its authors described it as "the first step in the practical use of this method for the construction of a five-year plan."[27] Subsequently, the table was used to work out twenty variants of the plan. The two "which best correspond to the political and economic tasks of the forthcoming five-year plan and also to the material and labor resources which will be available"[28] were apparently presented to the political leaders for their consideration. There are indications that the political leaders were not delighted with either of the variants; and little is now being said about the importance of the role played by input-output methods in the construction of the current five-year plan.

Work and experimentation with regional input-output tables at the republic level have been reported. Such activity has been going on in the Baltic republics for the past three or four years and was also involved, it is said, in some of these republics' work on constructing the five-year plan.[29]

All in all, in the application of mathematics to economics, Soviet economists and officials are clearly still at the stage of research and experiments. The hopes and expectations that some of the new mathematical methods and computer techniques would by now be introduced into planning management practice have (except for some operations research type of work) been largely unrealized.[30]

Why the Delay?

The simplest answer to the question "why the delay?" is perhaps also the most appropriate. To expect a fairly rapid introduction of the new mathematical methods for use on a scale greater than firm or small regional operations research type problems was ill-founded from the start. It takes a great deal of time to solve all the difficulties inherent in, and to make all the preparations necessary for, the shift of a large, complex economy to a new system of data reporting and processing, management decision-making, and plan construction and control. The Director of the Central Mathematical Economics Institute of the U.S.S.R. Academy of Sciences put the matter well recently. Everyone accepts the fact that it takes five to eight years to design and produce a new airplane, he said; then why be so impatient about the introduction of mathematical methods in economics? "Is the control of the economy any less complex than the flying of an airplane?"[31]

It was commonly thought that, once the Soviet economists worked out the theoretical mathematical problems, the introduction of mathematical methods in planning would proceed rapidly. Clearly, the Soviet economists have not yet worked out all the theoretical problems; but even if they had, there would still be an interminable number of decisions to be made and details laboriously to be worked out, before the new methods could be put into practice. One Soviet source has stated:

> It must be said that the mathematical capabilities have surpassed the capabilities of collecting and organizing economic information. This is possibly the reason why electronic computers have until now been used to solve relatively few problems of economic planning.[32]

The computerizing of the statistical system does not involve merely the substitution of the computer for the adding machine. New ways of reporting data must be devised before the data can be handled by computers. If there is to be a unified computer network, the data classifications must be standardized for all reporting data. Such standardization requires a complete overhaul of all existing accounting systems; it takes time to do this.

Computers—the basic "hardware"—must be designed and built (a time-consuming matter in itself), and the accompanying "software"—programs for the computers—must be devised. There is, incidentally, much complaint in the Soviet Union about the situation in regard to such "software."[33] In the absence of a supply of prepared computer programs, it often takes longer to handle a problem on the computer than by the old methods. Of course, once the program is worked out, it can be used over and over again (although the development of new equipment, which is rapid in the computer field, might require new programs or the adjustment of existing ones). Thus the production of computer "software" is essentially just as important a capital investment as the production of computer "hardware."

Even after the theoretical problems have been solved, the flow of data put under control, and computers made sufficiently available, it is still necessary to test the system in practice, alongside of the existing system. "It is wrong to suggest that first the scientific design of the system should be finished and then at once it should be introduced. The development of a system is unthinkable without much experimental testing on real economic objects which typify the situation in which it will be used."[34] Testing, too, takes time.

To work out the theoretical problems, to devise and institute new systems for the collection and organization of economic information, to produce computer "hardware" and "software," and then to conduct full-scale experimental tests of the system requires a massive investment of resources. Although the regime has been paying increasing attention to these matters and the indications are that even more resources have been devoted to them, the evidence suggests that the effort being made remains insufficient and even insignificant when compared with the magnitude of the task.[35] Insufficient resources may be an important factor in the slow pace of actual introduction of mathematical methods until now, and if the level of effort is not substantially increased, a slow pace in the future can be expected.

Any procedural and organizational change of the magnitude contemplated by many of the Soviet economists is bound to run into resistance from an entrenched bureaucracy. The political leaders must be firmly convinced of the necessity and the efficacy of the proposed change and must pursue it forcefully if it is to be accomplished. The rise to leadership of a man like Kosygin, who appears to understand well the present problems of the economy and the need for change, may increase the pace somewhat.

The foregoing has dealt with general reasons for the slow pace of the introduction of mathematics and computers into actual practice. For each mathematical method, there have in addition been specific difficulties which came to light as the method was experimented with. Again, the experience of the input-output method, from which so much was expected, is of particular interest.

The first major problem that arose in experiments with input-output methods concerned shortcomings of the available data. It was quickly seen, when work on the 1959 table began, that the data available through regular statistical channels were not usable. One problem, mentioned earlier, was the lack of standardized classifications for production materials. Organizations used different terms to describe the same material and grouped materials in different ways within different classifications (an interesting aspect of "departmental barriers" in the Soviet economy). A second, very important problem arose from the existence of multiproduct firms which produce not one single product or a homogeneous product group, but a heterogeneous group of products. The data they reported on their flow of inputs and outputs did not represent the physical relationships between the inputs and outputs of individual ("pure") product groups as required by the input-output method.

Because the regularly reported statistical data were not usable, a special survey covering 20 per cent of industrial and construction firms had to be conducted. And even then the accounting data regularly prepared by the firms for similar reasons could not be used to fill out the survey questionnaire, and the firms had to make special "economic calculations."[36] Even these extra efforts did not totally eliminate output heterogeneity (nor apparently lack of standardization). Those preparing the table had to employ further means to "purify" the sectors—that is, to remove nonsector outputs and inputs—so that the flows and coefficients would represent technological relationships between the individual product groups indicated by the sector titles.[37]

The purification of sectors for the planning model created a planning-administrative conflict. Because the targets in a Soviet plan are addressed to administrative organizations, who are then responsible for their fulfillment, the data in the input-output plan had to be laboriously converted back to the administrators' categories. In this regard, the table in physical units was superior to the one in value terms because it was closer to the administrative categories, but its sectors also did not fully coincide with administrative categories and its lack of full coverage of the set of commodities decreased its accuracy and thus weakened its usefulness.[38]

Finally, there were the problems associated with the shortcomings of the input norms. It should be emphasized that input norms are not derived from the input-output methods, but rather the reverse. The input-output method must utilize an already prepared set of direct input norms (and final outputs) in order to derive a set of direct and indirect input norms and a set of total outputs. Since the beginning of centralized planning, Soviet planners have had trouble constructing input norms, to such an extent that the economic literature has been filled with criticism of their poor quality. The significant point is that the unreliability of input norms may have been more damaging to the input-output experiment than it ever was to the material-balances method. For in the input-output approach, the norms are out in the open, at the center, and the whole system depends on how good they are. If they are accurate, the direct and indirect input coefficients are meaningful, and the total outputs are meaningful (given the appropriateness of the set of final demands). But if the norms are inaccurate, the results are not meaningful.[39]

In the case of the material-balances method, on the other hand, the input norms do not appear explicitly at the center; they are imbedded in the input requests sent up by the enterprises. Moreover, and most significantly, the material-balances method works through administrative processes as well as mathematical ones. That is, it can improve the internal consistency of the plan by applying pressure on producing organizations, in which process the input norms are "tightened." The norms, in effect, respond to the needs of the plan.

Altogether, it is quite possible that the large number of input norms involved,[40] their traditionally suspected low quality, their openness at the center of the plan, and the dependence of the entire system on their accuracy may have contributed to the opposition to introducing input-output methods into Soviet planning practice.

Science and Ideology in the New Economics

Because it has chronicled disappointed hopes, the tone
of this chapter has been quite negative. But despite little
accomplishment in actual planning practice, the mathematical revolution has had a profound impact on Soviet
economic thinking—an impact that is bound to have a
telling influence on the future course of the Soviet economy. For one thing, the woeful state of the economic
information system has been fully exposed, largely as a
result of the mathematical economists' need for accurate
and meaningful data. Though the system's deficiencies
have not posed insurmountable handicaps in the past,
growth of the economy makes such shortcomings increasingly serious. Exposure has focused attention on this
problem; improvement, with computerization, should in
time be forthcoming.

Even greater in importance has been the realization,
beginning as a trickle and developing into a flood, that
the new mathematical methods and computer techniques
could not simply be grafted on to old information systems
and old planning methods.[41] Just as the steam engine was
not merely a high-powered horse, so the computer is not
merely a high-powered adding machine. Its effective use
requires new approaches to economic questions. This has
led mathematical economists to rethink the entire field
of economic planning and organization.

The basic form of a mathematical approach to an
economic problem is to seek, through the logic of mathematics, from all the possible arrangements of given (limited) *means,* that arrangement which will achieve as high
a level as possible of desired *ends* (or, if the level of the
ends is given, to achieve them through a minimum ex-
penditure of means). This emphasizes the problem of

choice and the methods of making choices which lie at the core of the science of economics as it has been developed in the West over the course of the past century. Yet in the past, these have not been prominent features of Soviet economics. The mathematical approach also focuses attention on the question of ends—ends at all levels, from over-all economy-wide objectives to success indicators of the basic producing units, the firms.

Soviet economists face serious difficulties with respect to objectives of the firm. For example, after mathematical methods had shown how ton-kilometers expended in truck transport of freight could be minimized, the program was not widely adopted. It turned out that because the performance of trucking firms was measured in terms of ton-kilometers "produced," not "saved," truckers wanted to maximize rather than minimize their ton-kilometers.[42] Situations like this have led mathematical economists to pay attention to the meaningfulness of the measure of performance established for economic units. They have given strong support to the use of profit as the success indicator. For as long as the prices used are meaningful measures of relative values, profit, they recognize, is the most meaningful measure of performance in terms of the logic of the economic problem.

Concern with concepts of profits and prices has raised fundamental problems of ideology for Soviet mathematical economists. When open interest in the new science of mathematical economics emerged in 1956, the relationship of the sciences to Marxism was an immediate issue. The fact that mathematics was much used in bourgeois economics placed it under a cloud. This was countered in several ways. Such arguments as "Marx used mathematics" and "Marx said science can be perfected only if it makes use of mathematics" were advanced. Russian parentage for certain mathematical methods—linear programing and

input-output—was claimed. (The first was to a certain
extent true; the second, a gross overstatement.)[43] It was
argued that mathematics was merely a tool, devoid of
any ideology, which could be used for good purposes (by
Marxist economists) or for bad purposes (by bourgeois
economists). Nemchinov drew an analogy (subsequently
much quoted) between mathematics as a tool for economists and millstones as a tool for millers: "If a miller
grinds weeds, he will not get flour; but even if he starts
with the finest wheat, he will not get flour unless he uses
the millstones to grind it."[44]

The issue of the general acceptability of mathematics
was soon settled, but debate continued about the specific
acceptability of Kantorovich's "objectively-determined
evaluators." As noted above, these evaluators had all the
attributes of Western prices and, of course, Western economic theory was anathema to Marxist value theory. A
group of defenders tried to demonstrate the essential labor
content of the evaluators in an attempt to relate them to
the labor theory of value. Other defenders contended that
whereas bourgeois prices emerge from uncontrolled competition in the anarchy of the market, Kantorovich's evaluators emerge from a controlled interplay of mathemati-
cal-economic relationships within the computer. At a
meeting of economists called in March 1965 to discuss
the merits of the work of Kantorovich, Novozhilov, and
Nemchinov in light of their nominations for Lenin Prizes,
M. Golanskii attacked the ideological controversy in the
following manner:

> The basic accusation directed at the nominees consists of
> the question: Do the objective evaluators constitute value?
> This is scholasticism. Value and the objectively deter-
> mined evaluators are completely different and incommen-

surable things. Value is a category of political economy, and the objective evaluators constitute an algorithmic formula for the calculation of equilibrium prices in an optimum plan.[45]

In George Fischer's terminology,[46] what Nemchinov and Golanskii were saying was that mathematical methods and the economic levers they generate should not be considered elements of formal Marxist ideology, but rather that these matters should be "secularized." This is precisely what the political leaders have done by their award of Lenin Prizes to the three mathematical economists and by other acts and pronouncements.

The position of the particular concept of "marginalism," however, is not as yet clearly defined. Those who oppose basing official prices on marginal cost (which is both the Western marginalist solution and the linear-programming solution)—and they are still in the majority—do not rely solely on ideological arguments. They say such a step would result in the level of a price being set at the cost of the least efficient firm in the industry (which, while not precisely correct in theory, would in practice, undoubtedly occur). This would mean that all other firms in the industry could easily earn profits. And this touches the crux of the problem the Soviet economists and officials now face in their debates on price theory and price formation. In past and current Soviet practice, price has been a category of cost accounting, used primarily to control the fulfillment of the established plan, thus the fear of a price set at such a level that all firms easily earn profits.[47] Price has not been used to achieve efficiency in economic decision-making. Kantorovich's evaluators perform the latter function. The opposition to their use is, therefore, related to the maintenance of the present organization of Soviet planning and control. If economic reform and decentralization continue to advance, requiring more meaningful decision-making at lower levels, the acceptance of prices based on marginal cost will increase.

Ideology also becomes involved in the relationship between the use of mathematical methods in economics and that aspect of the exclusive rule of Party leaders which, in economic terminology, might be called the dominance of the preferences of the political leaders. The moment mathematical economists with their concern for means and ends, and for getting the most in the way of ends out of their means, think in terms of models encompassing the entire economy, they must think in terms of the over-all objective of the economy. They must do so, in fact, even when thinking about optimizing at the level of the firm. For optimizing the behavior of the firm will be rational only if the success indicator (criterion of optimality) of the firm is consistent with the over-all objective of the economy. Inevitably, then, the mathematical economist focuses on the economy's objective, its criterion of optimality.

What he wants and needs is a clear, single criterion of optimality or one in which various objectives are combined by means of clear quantitative weights (an "objective function").[48] The trouble, in an economic system where the political leaders' preferences as to ends are controlling, is that political leaders invariably do not supply such an "objective function." Political leaders do, of course, proclaim objectives, but not with the mathematical precision required for constructing a usable system of ends. Stalin's "economic laws" were statements of objectives, but they had no quantitative definiteness.[49]

The root of the problem lies in the difference between the exact, logical procedures of the mathematician and the inexact procedures of the political leader. Political leaders must operate and make decisions in the real world, where knowledge of the status, mode of operation, and interaction of socioeconomic phenomena is somewhat less than perfect. They may well be uneasy at the idea of an objective function, especially if the communication of their objective function to the planners is the only role they are to have in the process of constructing the over-all economic plan. The position taken at a 1962 conference by Boiarskii, a Soviet economist who is mathematically knowledgeable but conservative, is instructive. He was willing to accept the use of an objective function in the solution of economy-wide problems as long as a number of variants of the plan were constructed, from which the political leaders could choose the one they preferred.[50] Such choice would give the political leaders the opportunity to correct the effects produced by an objective function which did not truly reflect their objectives.

The specific content of a criterion of optimality for the economy has been a very popular subject in the Soviet economic literature in the last few years. It should be emphasized that there is no intrinsic conflict between the use of mathematical methods in economics and the dominance of political leaders' preferences in the objective function for the economy, whatever these preferences might be. Mathematical methods such as linear programming and input-output are only instrumental. They cannot produce ends and do not imply them. Ends must be established from the "outside," and whether they are determined by "democratic" or "dictatorial" procedures, in the economic marketplace or in the political arena, mathematical methods can be used to bring about their accomplishment.[51]

It is sometimes thought that mathematical economics is by its nature consumptionist—that is, favoring a larger share of the national product being comprised of consumption rather than investment (and defense) goods.

Yet the Soviet mathematical economists did not storm the Kremlin waving consumptionist banners the instant their computers were plugged in.[52] As some of them have declared:

> The relationship between the tendency to satisfy needs in the present and the tendency to satisfy needs in the future . . . must be determined by the decisions of the competent organs. . . .[53]

and:

> These decisions [regarding relative shares of consumption and investment] will inevitably change depending on the concrete socioeconomic situation in the country, and those economic goals which society will place before the national economy at each stage of the formation of the material-technical base of communism.[54]

Since the September 1965 Plenum, however, at which Kosygin emphasized the importance of consumption, articles on optimality increasingly have made the level of consumption dominant in the criterion of optimality. In fact, the two authors of the above quotations have recently changed their tone, asserting that consumption is the key criterion of optimality. Moreover, one of them (Dadaian) claims that "specialists on economic and mathematical methods have no differences" on this matter.[55] But, although mathematical economists generally now use consumption as *the* criterion of optimality in the models they build, this does not reflect either a necessary condition of using mathematical methods or a substitution of their preferences for those of the political leaders, but rather a shift in the leaders' value system (though some of the mathematical economists have gone further in this direction than have the political leaders).

The increasing importance of consumption in Soviet

economics raises another issue. How can the goods desired by consumers be put into an economy-wide objective
function? It is theoretically possible for the central planners to accomplish this through elaboration of "consumption norms." In past Soviet experience, however, this has
not been very successful or popular. Solving the problem,
then, will require either intense study (by means of econometrics) of consumer behavior—linking consumer demand to such variables as price, income, tastes, and social
status—or an increasing substitution of direct market relations for planning. If adoption of such measures constitutes an erosion of ideology, the erosion is a result not
of the introduction of mathematical methods but of other
forces at work in the society, including the level of economic development that has been attained, which have
led the political leaders to emphasize consumption more
and more.

A final point in regard to the effect on ideology of
the use of mathematical methods in economics concerns
the question of centralization versus decentralization.[56]
An earlier view that mathematics and computers operating
in a huge cybernetic system might prove to be the salva-
tion of centralization is no longer held in the Soviet Union
by many mathematical economists.[57] With his concentra-
tion on optimizing behavior, which requires the making
of choices from available alternatives at all levels of decision-making, the Soviet mathematical economist is con-
fronted by the enormity of the magnitude of detailed
knowledge involved. For this reason, he has become convinced of the impossibility of planning "every last bolt"
from the center. The dominant approach now is that of
decentralized cybernetics. It calls for two things: the
construction of a central plan which will indicate the
desired "trajectory" of the economy, and the use of a
price-profit system at the enterprise level which will

perform the function of a feedback mechanism to keep the economy moving on the planned trajectory.[58]

In a fascinating article,[59] Nemchinov once gave a theoretical justification, in Marxist terms, for using prices and profits on such an expanded scale. Though the words are significantly different, Nemchinov's line of reasoning recalls the argument by Golanskii quoted above. In essence, Nemchinov said that prices in the decentralized cybernetic system will reflect the preferences embodied in the over-all optimum economic plan. This subordination of the "law of value" to the "law of planned, proportional development" will create a situation where optimization at the lower levels will lead to the total optimization of the system, and thereby the achievement of the goals of the Party (whose objectives, of course, have been reflected in the over-all plan).

Even though decentralized management may mean a higher level of economic goal achievement, for the Party apparatus it means less direct power over the economy. Furthermore, use of the market will not lead to smooth performance and smooth growth. The market will, as is its nature, have its ups and downs, its successes and failures, and its other infirmities. There is, then, bound to be agitation from Party officials (and others) for a return to the familiar ground of centralization. The Soviet political leaders will need strength to resist this.

They may find some strength in a new faith, which Soviet science seems to share with Soviet ideology—the faith in cybernetics. For, as Loren Graham noted earlier, cybernetics is compatible with Marxism because of the basic strain of optimism they both possess. Cybernetics already shows signs of becoming the "invisible hand" of socialist markets.[60] It may help the Soviet system endure the fluctuations and failures that are bound to arise at times when market mechanisms are used.[61]

Thanks to an ideologically blessed faith in cybernetics, Soviet economists need not see these vacillations as aimless gropings in what Marx considered the anarchy of the capitalist market. Instead, a new Soviet economics might well fuse science and ideology. By this means it could come to think of socialist market forces as responses to feedback mechanisms, always returning a straying economy to the optimal path set for it by the country's ruling group, its political leaders.

NOTES: SOCIOLOGY

1 Clifford Geertz, "Ideology as a Cultural System," in David E. Apter, ed., *Ideology and Discontent* (New York: The Free Press, 1964), p. 48. On science and ideology, see also the essay by Reinhard Bendix, in the same volume, and six other recent studies: Norman Birnbaum, "The Sociological Study of Ideology," *Current Sociology*, IX, 2 (1960), reissued in 1966 by Johnson Reprint Corporation (New York); William E. Connolly, *Political Science and Ideology* (New York: Atherton Press, 1967); Jürgen Habermas, "Knowledge and Interest," *Inquiry*, IX, 4 (Winter 1966); the editor's introduction and conclusion, in Kurt Lenk, ed., *Ideologie* (Neuwied: Luchterhand, 2d edition, 1964); Francis X. Sutton *et al.*, *The American Business Creed* (Cambridge, Mass.: Harvard University Press, 1956), Chapters 1, 15; and Edward Shils, "Ideology," in *International Encyclopedia of the Social Sciences* (forthcoming).

2 Daniel Bell argues this view in his essay, "Marxism-Leninism, A Doctrine on the Defensive—The 'End of Ideology' in the Soviet Union?" in Milorad M. Dratchkovitch, ed., *Marxist Ideology in the Contemporary World* (New York: Praeger, 1966), and also in *Slavic Review*, XXIV, 4 (December 1965). While the present essay uses the field of sociology to argue against a total dichotomy between science and Soviet ideology, an American historian of science takes issue with the same Western view in terms of the natural sciences. David Joravsky does so, sharply and impressively, in two recent comments: "Soviet Ideology," *Soviet Studies*, XVIII, 1 (July 1966); and "Ideology, Science, and the Party," *Problems of Communism*, XVI, 1 (January–February 1967).

3 Geertz, cited in note 1, p. 71.

4 René Ahlberg, *Die Entwicklung der empirischen Sozialforschung in der Sowjetunion*, Report No. 60 of the Osteuropa-Institut of the Free University of Berlin (Wiesbaden: Harrassowitz, 1964); Robert V. Allen, "Recent Soviet Literature in Sociology and Cultural Anthropology," *Quarterly Journal of the Library of Congress*, XXII, 3 (July 1965); Leopold Labedz, "The Soviet Attitude to Sociology," in Walter Z. Laqueur and George Lichtheim, eds., *The Soviet Cultural Scene* (New York: Praeger, 1958); Leopold Labedz (or unsigned), articles in *Survey* since the mid-1950s; Alex Simirenko, ed., *Soviet Sociology, Historical Antecedents and Current Appraisals* (Chicago: Quadrangle Books, 1966); Elizabeth Ann Weinberg, "Soviet Sociology, 1960–1963" (Cambridge: Massachusetts Institute of Technology, Center for International Studies, 1964), processed; as well as my *Science and Politics: The New Sociology in the Soviet Union* (Ithaca, N.Y.: Cornell University, Center for International Studies, 1964).

Not long ago, a Soviet scholar published a quantitative survey of items of social research published in the Soviet Union. The survey fails to make clear on what grounds items were counted. By the same token, it seems to cover far too many kinds of items (such as "human interest" stories in the general press). Still, the survey points to a swift rise in output: Anatoly A. Zvorykin, "A Structural Analysis of Publications in the Field of Social [Research] in the Soviet Union, 1960–1965," *Social Research*, XXXIII, 4 (Winter 1966).

5 The stress here on these major works leads me to omit mention of most lesser writings. As a rule they are briefer and offer less data. A sampling of these writings can be found in a quarterly translation journal, *Soviet Sociology*, produced in White Plains, N.Y., by the International Arts and Sciences Press. The editor, Stephen P. Dunn, refers to some of its translations in *American*

Sociological Review, XXX, 6 (December 1965), 945–946, and in *Slavic Review*, XXIV, 4 (December 1965), 750–751.

6 The quotation is from a résumé of readers' letters to *Pravda*, on a related matter. English translation in *Current Digest of the Soviet Press* (New York), XVII, 23 (June 30, 1965), 15. *Current Digest of the Soviet Press* is hereinafter cited as CDSP.

7 *Metodologicheskie problemy nauki* [The Methodological Problems of Science] (Moscow: "Nauka," 1964), p. 134. The best Soviet summary of the debate appears in Chapter 6 of Arkadi I. Verbin and Aleksei I. Furman, *Mesto istoricheskogo materializma v sisteme nauk* [The Place of Historical Materialism in the System of Sciences] (Moscow: Moscow University Press, 1965); for some source references, see also pp. 3-4. In the present volume, Professor De George speaks of the same debate in the context of philosophy; see pp. 65–67.

8 The meeting of the Soviet Sociological Association which elected Osipov as the new president is described in *Voprosy filosofii*, XX, 6 (1966); English translation in CDSP, XVIII, 32 (August 31, 1966). A report on each panel of the Leningrad meeting appears in *Filosofskie nauki*, IX, 3 (1966).

The Academy's organizational decisions on social research appear in *Vestnik Akademii Nauk SSSR* (Moscow), XXXV, 5 (May 1966), 15–17. The article cited is *Izvestiia*, "Dlia progressa nauki i truda" [For the Progress of Science and Labor], May 12, 1966, by A. M. Rumiantsev, T. Timofeiev, Iu. Shelpen; English translation in CDSP, XVIII, 16 (June 1, 1966), 36–37. On Academician Rumiantsev, some biographic data and conjectures can be found in *Problems of Communism*, 16, 1 (January–February 1967), 31.

9 Brezhnev's remarks on sociology appear in English translation in CDSP, XVIII, 13 (April 20, 1966), 8; Kosygin's remarks appear a week later, No. 14 (April 27, 1966), 18. Both were delivered at the Twenty-third Party Congress, Brezhnev's appearing originally in *Pravda*, March 30, 1966, pp. 2–9, and Kosygin's in *Pravda*, April 6, 1966, pp. 2–7.

10 *Izvestiia* published a long interview with Osipov (December 18, 1965, p. 6); English translation in CDSP, XVII, 51 (January 12, 1966), 16–17. On March 13, 1966, *Pravda*, p. 3, carried a piece by Vladimir N. Shubkin; English translation as "Problems and Prospects of Sociology," CDSP, XVIII, 11 (April 6, 1966), 4–5. One such statement, made by Boris A. Grushin, was printed on the front page of *Literaturnaia gazeta*, September 25, 1965; English translation as "Sociology and Sociologists," CDSP, XVII, 40 (November 27, 1965), 15–16. In part, Grushin said: "Sociology will not begin to produce tangible benefits for the theory and practice of social construction until it receives the

'lawful' organizational and material expression that every 'normal' science must have. The minimum should be the establishment of a Sociology Institute and at least one sociological journal which could head up the scientific research conducted at all levels, provide skilled guidance, coordinate work, etc. And it is necessary to start training specialized sociological cadres on an extensive scale. It would seem perfectly obvious that no science can exist without its scientists, yet in the entire country there is not one sociologist (!) who received special training in sociology under a full program at the present-day scientific level." A set of more recent comments by Soviet sociologists on the state of their field points to its new popularity, and to the risks involved in trying to do (or to promise) too much in too short a time: *Literaturnaia gazeta*, August 6, 1966; translated in CDSP, XVIII, 32 (August 31, 1966).

11 Academician Aleksandr D. Aleksandrov, "Slovo o sotsiologii" [A Word on Sociology], *Literaturnaia gazeta*, April 21, 1966, and S. Muroshov, "Student i obshchestvennie nauki" [The Student and the Social Sciences], *Komsomol'skaia Pravda* (Moscow), April 20, 1966; English translation of both items in CDSP, XVIII, 17 (May 18, 1966), 7 and 5, respectively. Vladimir Kantorovich, "Rodstvennaia nam nauka" [A Science Kindred to Us], *Literaturnaia gazeta*, May 5 and 14, 1966; English translation in CDSP, XVIII, 26 (July 20, 1966), 14–16.

In the fall of 1966, the main Soviet newspaper for intellectuals, *Literaturnaia gazeta*, launched a large section devoted to sociology (the issue of November 12, 1966). At about the same time, the country's leading intellectual journal, *Novyi mir*, published its first major essay by a sociologist, Igor S. Kon, "The Psychology of Prejudice," XLII, 9 (September 1966); English version, *Soviet Anthropology and Archeology*, V, 2 (Fall 1967).

12 Anatoly A. Zvorykin, *Nauka, proizvodstvo, trud* [Science, Production, Labor] (Moscow: "Nauka," 1965); V. N. Yermuratski, *et al.*, eds., *Kopanka 25 let spustia* [Kopanka 25 Years Later] (Moscow: "Nauka," 1965); Vladimir I. Razin, ed., *Stanovlenie kommunisticheskogo samoupravleniia* [The Development of Communist Self-Government] (Moscow: Moscow University Press, 1965); and E. N. Chesnokov *et al.*, eds., *Puti preodaleniia sushchestvennykh razlichii mezhdu umstvennym i fizicheskim trudom* [The Patterns of Overcoming Essential Differences Between Mental and Physical Labor] (Moscow: "Mysl'," 1965).

13 Ralf Dahrendorf, *Class and Class Conflict in Industrial Society* (Stanford: Stanford University Press, 1959), p. 313.

14 Lazarsfeld has made the point that quantitative social research is not quite as purely American in origin as many believe. He and also Oberschall cite the case of Germany, which took its

first steps in "sociology" in the early twentieth century. Soviet sociologists make the same point about their own country in the 1920s. They stress especially the studies of Academician Stanislav G. Strumilin. See Paul F. Lazarsfeld, "Preface to New Edition," in Marie Jahoda *et al.*, *Die Arbeitslosen von Marienthal* (Allensbach: Verlag für Demokopie, 1960); Anthony Oberschall, *Empirical Social Research in Germany, 1848–1914* (Paris: Mouton, 1965); and Stanislav G. Strumilin, *Izbrannye proizvedeniia, Problemy ekonomiki truda* [Selected Works, Problems of Labor Economics], III (Moscow: "Gospolitizdat," 1957).

15 The accounts are translated in CDSP, XVI, Nos. 46, 47, and 48 (December 1964). Allen Kassof commented on the early *Komsomol'skaia Pravda* polls in "Moscow Discovers Public Opinion Polls," *Problems of Communism*, X, 3 (May–June 1961). The methods used in these polls are sketched (by Boris A. Grushin) in G. K. Ashin *et al.*, eds., *Voprosy organizatskii i metodiki konkretno-sotsiologicheskikh issledovanii* [Questions of the Organization and Methods of Concrete Sociological Research] (Moscow: "Rosvuzizdat," 1963).

16 On leisure, see F. Gayle Durham, "The Use of Free Time by Young People in Soviet Society" (Cambridge: Massachusetts Institute of Technology, Center for International Studies, 1966, processed).

The Multinational Comparative Time Budget Research Project is perhaps the first major comparative study in which Soviet sociologists took part. Sponsored by the Vienna Center of the European Social Science Council and headed by Alexander Szalai of Hungary, the study is described in a special report in *American Behavioral Scientist*, X, 4 (December 1966). The report consists of a 31-page introduction and, under separate cover, a 61-page appendix with five series of tables.

17 Gennadi V. Osipov *et al.*, eds., *Rabochii klass i tekhnicheskii progress, Issledovanie izmenenii v sotsial'noi strukture rabochego klassa* [The Working Class and Technical Progress, A Study of Changes in the Social Structure of the Working Class] (Moscow: "Nauka," 1965).

18 Andrei G. Zdravomyslov and Vladimir A. Yadov, editors and senior authors, *Trud i razvitie lichnosti* [Work and Personality Development] (Leningrad: "Lenizdat," 1965). An article by Zdravomyslov and Yadov in *Voprosy filosofii*, XVIII, 4 (April 1964); English translation, "An Experiment in Concrete Research on Attitudes Toward Labor," in CDSP, XVI, 24 (July 8, 1964), 12–17. Article by Zdravomyslov and Yadov in Gennadi V. Osipov, ed., *Sotsiologiia v SSSR* [Sociology in the U.S.S.R.], II (Moscow: "Mysl," 1965).

19 Vladimir N. Shubkin *et al.*, *Kolichestvennye metody v sotsi-*

ologicheskikh issledovaniiakh [Quantitative Methods in Sociological Research], edited by Shubkin (Novosibirsk: Novosibirsk University Press, 1964) Part II, pp. 152–267. Article by Shubkin in *Voprosy filosofii*, XVIII, 8 (September 1964), 18–28. Article by Shubkin, "Molodezh' vstupaet v zhizn'," in *Voprosy filosofii*, XIX, 5 (May 1965); English translation as "Youth Enters Life" in CDSP, XVII, 30 (August 18, 1965), 3–9. Article by Shubkin in Nikolai V. Novikov, *et al.*, eds., *Sotsial'nye issledovaniia* [Social Research] (Moscow: "Nauka," 1965). Chapter VII of *Kolichestvennye metody v sotsiologii* [Quantitative Methods in Sociology], edited by Shubkin (Moscow: "Nauka," 1966), pp. 168–231.

The last of these items constitutes a revision and updating of the first item on the Novosibirsk study. But the 1966 version omits two kinds of data which made the earlier 1964 version stand out among recent Soviet studies: the text of the questionnaires used, and a full list of occupations and how each of them ranked among various groups of high school seniors (pp. 244–267, 1964 version).

In the wake of the Novosibirsk study, sociologists in Sverdlovsk have dealt with the same theme: Mikhail N. Rutkevich, ed., *Zhiznennyie plany molodezhi* [The Life Plans of (Soviet) Youth] (Sverdlovsk: Ural University Press, 1966).

Two more empirical studies are under way in Sverdlovsk. The first deals with the impact of education on the cultural life of workers. An interim report appears in M. T. Iovchuk and L. N. Kogan, "Changes in the Spiritual Life of Workers in the USSR" (in Russian), *Filosofskie nauki*, IX, 6 (1966).

A further Sverdlovsk study centers on the changing social composition of the Soviet intelligentsia. "Intelligentsia" is here limited to the most skilled, the "specialists" whose work as a rule calls for higher (or semi-professional secondary) education; as against past Soviet usage, this definition of intelligentsia leaves out the less skilled "non-specialists" of the "gray collar" stratum. The intelligentsia study is headed by a leading sociologist who serves as the Dean of the Philosophy Faculty of Ural University: Mikhail N. Rutkevich. He reports on the study in two collections of papers—Volume I of *Sotsiologiia v SSSR*, cited in note 18, and *Izmenenie sotsialnoi struktury sovetskogo obshchestva* [Changes in the Social Structure of Soviet Society] (Moscow: "Znanie," 1966)—and in two articles, in *Filosofskie nauki*, IX, 4 (1966), and *Voprosy filosofii*, XXI, 6 (1967).

Like themes are touched upon in yet another collection of technical essays (such as the one with the main Shubkin study, cited in this note), put out by sociologists in Novosibirsk: R. V. Ryvkina, ed., *Sotsiologicheskie issledovaniia, Voprosy metodo-*

logii i metodiki [Social Research, Problems of Methodology and Methods] (Novosibirsk University Press, 1966).

20 Zdravomyslov and Yadov, in *Sotsiologiia v SSSR*, II (cited in note 18), p. 208.

21 Shubkin, CDSP translation of "Youth Enters Life," cited in note 19, pp. 4–5.

22 Shubkin, p. 6.

23 Shubkin, p. 7.

24 Shubkin, p. 9.

25 James S. Coleman, "Relational Analysis, The Study of Social Organizations with Survey Methods," in Amitai Etzioni, ed., *Complex Organizations* (New York: Holt, Rinehart & Winston, 1961).

26 Talcott Parsons, "An American Impression of Sociology in the Soviet Union," *American Sociological Review*, XXX, 1 (February 1965), 124. The same pattern is noted—and criticized—by a Soviet writer (cited in note 11): Vladimir Kantorovich, writing in *Literaturnaia gazeta*, May 14, 1966; English translation in CDSP, XVIII, 26 (July 20, 1966), 14–16.

27 Today quite a few Soviet sociologists follow Western work in the field much more closely (and expect graduate students to know it far more fully) than is true the other way around. Such Soviet scholars now tend to speak out sharply against their own colleagues, too, when they misstate in print some major fact about Western writings. See, for example, V. A. Karpushin of the Philosophy Institute, and Igor S. Kon, of Leningrad University, in *Voprosy filosofii*, XIX, 1 (January 1965), 173–175; English translation in CDSP, XVII, 19 (June 2, 1965), 12–13. Here Karpushin and Kon review a Soviet volume of essays on the Fifth World Congress of Sociology: *Marksistskaia i burzhuaznaia sotsiologiia segodnia* [Marxist and Bourgeois Sociology Today], F. V. Konstantinov, *et al.*, eds. (Moscow: "Nauka," 1964). See also two comments made by Yuri A. Zamoshkin in a recent issue of *Voprosy filosofii*, XX, 3 (March 1966), 160–161, as well as pp. 145–147.

A major review of the 1964 volume on the Fifth World Congress of Sociology, just cited, appears in *Voprosy filosofii*, XX, 2 (1966). In part quite critical in tone, the review stems from a widely known younger economist, A. G. Aganbegian. For a translation of the review, see *Soviet Sociology*, V, 3 (Winter 1966–1967).

28 Gennadi V. Osipov, *Sovremennaia burzhuaznaia sotsiologiia* [Contemporary Bourgeois Sociology] (Moscow: "Nauka," 1964); G. M. Andreeva, *Sovremennaia burzhuaznaia empiricheskaia sotsiologiia* [Contemporary Bourgeois Empirical Sociology] (Moscow: "Mysl'," 1965).

Soviet scholars have begun to publish translations of leading present-day Western sociologists. Osipov has edited a translation of *Sociology Today*, the major 1959 collection of essays sponsored by the American Sociological Association: *Sotsiologia segodnia* (Moscow: "Progress," 1965). The translation runs to 684 pages. Osipov also edited a translation of work in mathematical sociology, *Matematicheskie metody v sovremennoi burzhuaznoi sotsiologii* [Mathematical Methods in Contemporary Bourgeois Sociology] (Moscow: "Progress," 1966). The 400-page volume contains essays by Georg Karlsson and Herbert Simon on social models, by Louis Guttman on scale analysis, and by Paul F. Lazarsfeld on latent-structure analysis.

Some translations of Soviet work are now coming out in English. Tavistock Publications of London (together with Barnes and Noble of New York) have begun to issue a translation series of four volumes. The series is called *Studies of Soviet Society*. A first volume, already in print, deals with *Industry and Labour in the U.S.S.R.* (1966). The International Arts and Sciences Press of White Plains, N.Y., has made known its plans to publish another translation volume: *Sociology in the Soviet Union*, edited by Osipov.

About twenty of the Soviet papers submitted to the Sixth World Congress of Sociology (Evian, France, September 1966) can be found in French translation, *La sociologie en U.R.S.S.* (Moscow: "Progress," 1966).

29 Karl Mannheim, *Ideology and Utopia* (New York: Harcourt, Brace, 1946), pp. 68–69n.

30 Andreeva, cited in note 28, pp. 186, 187.

31 In Russian sociology, some of these ways of thinking go back to an early, pre-Marxian phase. This can be seen in Pitirim A. Sorokin, "Russian Sociology in the Twentieth Century," *Publications of the American Sociological Society*, XXI (1926); Max Laserson, "Russian Sociology," in Georges Gurvitch and Wilbert E. Moore, eds., *Twentieth Century Sociology* (New York: Philosophical Library, 1945); and Klaus von Behme, *Politische Soziologie im zaristischen Russland* (Wiesbaden: Harrassowitz, 1965).

32 Robert K. Merton, "Social Conflict over Styles of Sociological Work," *Transactions of the Fourth World Congress of Sociology*, III (Louvain: International Sociological Association, 1961), p. 31. A range of negative American reactions can be seen in Vladimir C. Nahirny, "Soviet Criticism of Western Sociology," *American Catholic Sociological Review*, XIX, 3 (October 1958); Lewis S. Feuer, "Meeting the [Soviet] Philosophers," *Survey*, No. 51 (April 1964); Allen Kassof, "American Sociology through Soviet Eyes," *American Sociological Review*, XXX, 1 (February

1965); and Paul Hollander, "The Dilemmas of Soviet Sociology," *Problems of Communism*, XIV, 6 (November–December 1965).

33 Jerzy J. Wiatr, "Political Sociology in Eastern Europe, A Trend Report and Bibliography," *Current Sociology*, XIII, 2 (1965).

34 Wiatr, cited in note 33, pp. 58–59.

35 Wiatr, pp. 64–65.

36 Wiatr, p. 66.

37 Adam Schaff, *Marxismus und das menschliche Individuum* (Vienna: Europa Verlag, 1965), pp. 268–269.

38 Aleksandra Jasinska, "Systematizing the Teachings of Karl Marx," *Polish Sociological Bulletin* (in English), XIII, 1 (1966), 11. Two new surveys of research, on key aspects of social stratification and mobility, also bring out the kind of work that Polish sociologists now do while their Soviet colleagues do not: Aleksander Matejko, "Status Incongruence in the Polish Intelligentsia," *Social Research*, XXXIII, 4 (Winter 1966): Wlodzimierz Wesolowski, on classes and strata, in *Sociologie du Travail*, IX, 2 (1967).

39 Galina M. Andreeva, "The Formation of New Representatives of the Intelligentsia and Leaders in the Course of the Building of Socialism," *Transactions of the Fifth World Congress of Sociology*, III (Louvain: International Sociological Association, 1964), pp. 233–234n. Severyn Bialer sums up Andreeva's argument and comments on it in his "Soviet Political Elite: Concept, Sample, Case Study," unpublished Ph.D. dissertation in political science, Columbia University, 1966, pp. 1–3.

40 Osipov *et al.*, cited in note 17.

41 Early in the book on the Gorky study, a footnote quotes with approval a 1965 plea in the Party's main theoretical journal. There Soviet philosophers are urged to give up the idea that the concept of alienation has been forced upon them by foreign critics, and to "develop" the Marxist concept "on the basis of contemporary materials" (p. 10n). The quoted passage appeared in *Kommunist*, XLII, 5 (March 1965), 63.

42 Osipov *et al.*, cited in note 17, p. 43.

43 Osipov *et al.*, p. 44.

44 Osipov *et al.*, p. 42.

45 Osipov *et al.*, p. 15.

46 This proposition is not unlike a finding of an American work on the same subject: Robert Blauner, *Alienation and Freedom* (Chicago: University of Chicago Press, 1962). The Soviet book refers to Blauner's work but not to this finding.

47 The fate of division of labor under communism was disputed sharply among Soviet social scientists in the early 1960s. I touch on these disputes in *Science and Politics* (cited in note 4), pp. 64–65. On collectivism as an important element in mak-

ing diversified work both possible and highly satisfying, the book on the Gorky study often refers to Yuri N. Davydov, *Trud i svoboda* [Work and Freedom] (Moscow: "Vysshaia shkola," 1962; New York: Johnson Reprint Corp., 1968); a German translation appeared under the title *Freiheit und Entfremdung* ([East] Berlin: Deutscher Verlag der Wissenschaften, 1964).

48 Osipov *et al.*, cited in note 17, p. 16.

49 Lewis A. Coser, "Karl Marx and Contemporary Sociology," in his *Continuities in the Study of Social Conflict* (New York: The Free Press, 1967).

50 Osipov *et al.*, cited in note 17, p. 296.

51 G. M. Andreeva and E. P. Nikitin, "The Method of Explanation in Sociology" (in Russian), in G. V. Osipov, ed., *Sotsiologiia v SSSR* [Sociology in the U.S.S.R.], I (Moscow: "Mysl'," 1965); and two essays by V. A. Yadov (both in Russian): "On the Establishment of Facts in Concrete Sociological Research," *Filosofskie nauki*, IX, 5 (1966), 28–38; and "The Role of Methodology in Defining the Methods and Technique of Concrete Sociological Research," *Voprosy filosofii*, XX, 10 (1966), 27–37.

The methodology of recent work is surveyed by Susan Gross Solomon in "Theory and Research in Soviet Social Inquiry," unpublished M.A. thesis, Columbia University, 1967.

52 The present trends of sociology in the Soviet Union (and in East Central Europe as well) are the subject of four new Western studies. The largest of them comes from a sociologist in Western Germany: Gabor Kiss, *Gibt es eine "marxistische" Soziologie?* (Cologne: Westdeutscher Verlag, 1966). The other studies are Leopold Labedz, "Sociology and Social Change" (Part One), *Survey*, No. 60 (July 1966); Janina Markiewicz-Lagneau, "Les problèmes de mobilité sociale en U.R.S.S.," *Cahiers du monde Russe et Soviétique*, VII, 2 (1966); and John S. Shippee, "Empirical Sociology in the East-European Communist Party-States" (Stanford: Stanford University, Stanford Studies of the Communist System, November 1966, processed). Useful notes on new social research can be found in the quarterly *Information Supplement* of *Soviet Studies*. On "realistic" ideology, see the Joravsky essay in *Soviet Studies*, cited in note 2.

53 One work especially cannot be assigned to either orientation. It fits somewhere in between. I have in mind a book by a leading Soviet specialist on the family, a member of the Leningrad Branch of the U.S.S.R. Academy of Sciences: Anatoli G. Kharchev, *Brak i semia v SSSR* [Marriage and Family in the U.S.S.R.] (Moscow: "Mysl'," 1964). Nearly half of this book

deals with family life not in the Soviet Union but in the West.

54 Kurt H. Wolff, "The Sociology of Knowledge and Sociological Theory," in Llewellyn Gross, ed., *Symposium on Sociological Theory* (New York: Harper and Row, 1959), see especially pp. 580–582.

55 Not long ago this point was made by Peter L. Berger in "Reification and the Sociological Critique of Consciousness" (with Stanley Pullberg) and "Response," in *New Left Review*, No. 35 (January–February 1966). Other recent comments along similar lines can be found in Barrington Moore, Jr., "Tolerance and the Scientific Outlook," in Robert Paul Wolff, *et al., A Critique of Pure Tolerance* (Boston: Beacon Press, 1965); Lewis A. Coser, *Men of Ideas* (New York: The Free Press, 1965), pp. 358–361; Jürgen Habermas, *Theorie und Praxis* (Neuwied: Luchterhand, 1963), pp. 215–230; John Horton, "The Dehumanization of Anomie and Alienation," *British Journal of Sociology*, XIV, 4 (December 1964); Andrew Gunder Frank, "Functionalism, Dialectics and Synthetics," *Science and Society*, XXX, 2 (Spring 1966). On the other side, Etzioni has argued anew that macroanalysis should be added to the mainstream of American sociology—and can be done best there. See Amitai Etzioni, "Social Analysis as a Sociological Vocation," *American Journal of Sociology*, LXX, 5 (March 1965).

A Soviet scholar has written a most interesting piece on "critical sociology." But he deals only with the American scene and says nothing of his own country. Igor S. Kon, "Weltanschauung und 'kritische Soziologie' in der USA," *Deutsche Zeitschrift für Philosophie* ([East] Berlin), XIV, 2 (1966).

NOTES: PHILOSOPHY

1 My use of "ideology" here is consistent with that suggested by Professor Fischer in the preceding chapter. For a more detailed study of the nature of Soviet ideology and its relation to philosophy, see Chapter XII of my *Patterns of Soviet Thought* (Ann Arbor: University of Michigan Press, 1966).
2 V. I. Lenin, *Polnoe sobranie sochinenii* [Collected Works] (Moscow: Gospolitizdat, 1958–1965), 5th ed., Vol. XX, p. 84, which gives a quote from Engels as well; *Filosofskaia entsiklopediia* (Moscow: Gosudarstvennoe nauchnoe izdatel'stvo, 1962), II, 362, is a contemporary reiteration of the position.
3 See, for example, H. B. Acton, *The Illusion of the Epoch* (London: Cohen & West Ltd., 1955), pp. 141–179; Sidney Hook, *Marx and the Marxists* (Princeton, N.J.: Van Nostrand, 1955), pp. 35ff.; T. J. Blakeley, *Soviet Scholasticism* (Dordrecht: Reidel, 1961), pp. 39–71.

4 The same *is* true of such other basic components of Marxism-Leninism as the collectivist view of man which it preaches. Collectivism and the individualism to which it is opposed are not objective, descriptive views of man, but evaluative views which describe objective aspects of man from a certain position or focus. Even such basic components of dialectical materialism as the doctrine of materialism have evaluative aspects. In the doctrine of materialism, for instance, there is involved not only a factual claim about the ultimate nature of reality but also a commitment to this world and a rejection of religion and religious values. Thus, when Engels defends materialism against Dühring, his arguments consist most frequently of showing that Dühring's position leads to the postulation of God, a postulation which Engels rejects out of hand. Similarly in the *Filosofskie tetradi* [Philosophical Notebooks], Lenin, in commenting on Hegel's *Logic,* claims in the same breath that Hegel scarcely says a word about God, and that "in this *most idealistic* of Hegel's works there is the *least* idealism and the *most materialism*" (*Polnoe sobranie sochinenii,* XXIX, 215).

5 These include the following: matter is all that exists; reality is essentially dialectical; the triumph of communism is inevitable; the aims of communism coincide with the aims of working mankind; and the Party is the vanguard of mankind and will lead it to communism.

6 T. J. Blakeley applies this distinction to Soviet philosophers working in the field of the theory of knowledge. See his "Is Epistemology Possible in Diamat?" *Studies in Soviet Thought,* II, 1 (March 1962), 95–103.

7 *Filosofskaia entsiklopediia,* II, p. 353.

8 *Ibid.,* pp. 363–364; A. Verbin and A. Furman, *Mesto istoricheskogo materializma v sisteme nauk* [The Place of Historical Materialism in the Sciences] (Moscow: Moscow University Press, 1965), p. 15.

9 See especially Karl R. Popper, *The Poverty of Historicism* (Boston: Beacon Press, 1957), and Isaiah Berlin, *Historical Inevitability* (Oxford: Oxford University Press, 1955).

10 For a brief summary of some of the discussions which took place between 1956 and 1963, see *Marksistsko-leninskaia filosofiia i sotsiologiia v SSSR i evropeiskikh sotsialisticheskikh stranakh* [Marxist-Leninist Philosophy and Sociology in the U.S.S.R. and the Socialist Countries of Europe] (Moscow: "Nauka," 1965), pp. 262–289.

11 A. P. Butenko, "Dve mirovye sistemy i tendentsii ikh razvitiia" [Two World Systems and Their Development Tendencies], *Voprosy filosofii,* XIX, 9 (September 1965), 3–13.

12 Popper defines "historicism" as "an approach to the social sci-

ences which assumes that *historical* prediction is their principal aim, and which assumes that this aim is attainable by discovering the 'rhythms' or the 'patterns,' the 'laws' or the 'trends' that underlie the evolution of history." Cited in note 9, p. 3.

13 Joseph Stalin, *Dialectical and Historical Materialism* (New York: International Publishers, 1940). The other two laws of dialectics, as formulated by Engels, are the law of the transition from quantity to quality and vice versa, and the law of the interpenetration of opposites.

14 Joseph Stalin, *Marxism and Linguistics* (New York: International Publishers, 1951), pp. 27–28. For a fuller discussion on this and the preceding reference, see my *Patterns of Soviet Thought*, Chapter X.

15 For both sides of this debate, see "Aktual'nye problemy dialektiki" [Urgent Problems of the Dialectic], *Voprosy filosofii*, XIX, 10 (October 1965), 130–164. See also L. N. Suvorov, "O meste istoricheskogo materializma v strukture marksistsko-leninskoi filosofii" [The Place of Historical Materialism in the Structure of Marxist-Leninist Philosophy], *Voprosy filosofii*, XVIII, 8 (October 1964), 139–142.

16 See for example, V. S. Tiukhin, " 'Kletochka' otrazheniia i otrazhenie kak svoistvo vsei materii" [The "Primary Cell" of Reflection and Reflection Itself as a Property of All Matter], *Voprosy filosofii*, XVIII, 2 (February 1964), 25–34.

17 For a report of one such conference, see note 13 above.

18 Stalin, *Marxism and Linguistics*, p. 34.

19 V. Zh. Kelle and M. Ia. Koval'zon, "Kategorii istoricheskogo materializma" [Categories of Historical Materialism], *Voprosy filosofii*, X, 4 (April 1956), 18–32; V. P. Tugarinov, *Sootnoshenie kategorii istoricheskogo materializma* [The Interrelation of the Categories of Historical Materialism] (Leningrad: Leningrad University Press, 1958); *Kategorii istoricheskogo materializma* [The Categories of Historical Materialism] (Frunze: Kirghiz State University, 1959).

20 G. E. Glezerman, *O zakonakh obshchestvennogo razvitiia* (Moscow: "Gospolitizdat," 1960). Glezerman's book is available in English translation as *The Laws of Social Development* (Moscow: Foreign Languages Publishing House, 1962); *Dialektika razvitiia sotsialisticheskogo obshchestva* [The Dialectic of the Development of Socialist Society] (Moscow: Izdatel'stvo VPSh i AON, 1961); D. I. Chesnokov, *Istoricheskii materializm* [Historical Materialism] (2d ed., Moscow: "Mysl'," 1965).

21 A. E. Furman, "O predmete istoricheskogo materializma" [Concerning the Subject of Historical Materialism], *Filosofskie nauki*, VIII, 6 (1965), 85–90; M. S. Dzhunusov, "O vzaimosviazi osnovnykh nauchnykh poniatii istoricheskogo materializma" [Con-

cerning the Interconnection of the Principle Concepts in Historical Materialism], *Voprosy filosofii,* XIX, 7 (July 1965), 144–
146; M. Kammari, "Nekotorye voprosy teorii bazisa i nadstroiki"
[Several Questions on the Theory of Basis and Superstructure],
Kommunist, No. 10 (July 1956), pp. 42–58.

22 Furman, cited in note 21, p. 88.

23 V. P. Tugarinov, "O kategoriiakh 'obshchestvennoe bytie' i
'obshchestvennoe soznanie' " [Concerning the Categories "Social
Existence" and "Social Consciousness"], *Voprosy filosofii,* XII,
1 (January 1958), 15–26.

24 G. E. Glezerman, "K voprosu o poniatii 'obshechestvennoe
bytie' " [On the Question of the Concept of "Social Existence"],
Voprosy filosofii, XII, 5 (May 1958), 117–126.

25 *Marksistsko-leninskaia filosofiia i sotsiologiia v SSSR i evropeiskikh stranakh,* cited in note 10, p. 269.

26 *Filosofskaia entsiklopediia,* II, p. 358.

27 See my "A Bibliography of Soviet Ethics," and "Soviet Ethics
and Soviet Society," both in *Studies in Soviet Thought,* III, 1
(March 1963), 83–103, and IV, 3 (September 1964), 206–217,
respectively.

28 A. F. Shishkin, *Osnovy marksistskoi etiki* [The Bases of Marxist
Ethics] (Moscow: IMO, 1961), pp. 106–109.

29 Verbin and Furman, cited in note 8; P. N. Fedoseev, "Filosofskoe obosnovanie nauchnogo kommunizma" [The Philosophical
Basis of Scientific Communism], *Voprosy filosofii,* XVIII, 11
(November 1964), 3–14; D. I. Chesnokov, "Vzaimootnoshenie
obshchestvennykh nauk i mesto Nauchnogo kommunizma sredi
nikh" [The Interrelationship of the Social Sciences and the Place
of Scientific Communism among Them], *Voprosy filosofii,* XIX,
3 (March 1965), 20–31; *Nauchnyi kommunizm* [Scientific Com-
munism] (Moscow: Izdatel'stvo politicheskoi literatury, 1965).

30 See the chapter by George Fischer in this volume.

31 F. V. Konstantinov and V. Zh. Kelle, "Istoricheskii materializm—
marksistskaia sotsiologiia" [Historical Materialism is Marxist
Sociology], *Kommunist,* No. 1 (January 1965), p. 23.

32 D. M. Gvishiani, "Istoricheskii materializm i chastnye sotsiologicheskie issledovaniia" [Historical Materialism and Specific
Sociological Research], *Voprosy filosofii,* XIX, 5 (May 1965),
49, 56.

33 D. I. Chesnokov, "Predmet i zadachi istoricheskogo materializma" [The Subject and Tasks of Historical Materialism],
Filosofskie nauki, VIII, 1 (1965), 50.

34 Quoted in Verbin and Furman, cited in note 8, pp. 146, 155.

35 Chesnokov, cited in note 33, p. 50; Furman, cited in note 21,
pp. 85–90.

36 A. M. Gendin, "Rol' sotsial'nogo predvideniia i tseli v razvitii

sotsialisticheskogo obshchestva" [The Role of Social Foresight and Goals in the Development of Socialist Society], *Filosofskii nauki*, IX, 1 (1966), 24–31.

37 See, for instance, Stalin's *Problems of Leninism* (Moscow: Foreign Languages Publishing House, 1954), p. 778, where he speaks of the moral and political unity of Soviet society, the mutual friendship of the nationalities of the U.S.S.R. and Soviet patriotism as motive forces in the development of Soviet society. See also his discussion on linguistics.

38 Cf. the "Moral Code of the Builder of Communism" in the 1961 Party Program; see also the articles cited in note 27 above, and G. G. Filippov, "O roli moral'nogo faktora v sovremennom promyshlennom proizvodstve" [Concerning the Role of the Moral Factor in Contemporary Industrial Production], *Filosofskii nauki*, VIII, 5 (1965), 47–54.

39 See the discussions on communism and the division of labor in the following issues of *Voprosy filosofii*; XV, 11 (1961); XVI, 10 (1962); XVII, 3, 9, 11, 12 (1963); XVIII, 1, 6 (1964). On alienation see T. I. Oizerman, "Man and His Alienation" (in English), *Philosophy, Science, and Man* (Moscow: U.S.S.R. Academy of Sciences, 1963), pp. 99–107.

40 *Filosofskaia entsiklopediia*, II, p. 358.

41 G. E. Glezerman, "Dialektika ob'ektivnykh uslovii i sub'ektivnogo faktora v stroitel'stve kommunizma" [The Dialectic of Objective Conditions and Subjective Factors in Building Communism], *Voprosy filosofii*, XIX, 6 (June 1965), 13.

42 I. F. Balakina, "Probleme tsennostei—vnimanie issledovatelei" [Scholars Should Pay Attention to the Problem of Value], *Voprosy filosofii*, XIX, 9 (September 1965), 154.

43 P. N. Fedoseev, "Dialektika razvitiia sotsializma" [The Dialectic of the Development of Socialism], *Kommunist*, No. 14 (September 1965), pp. 28–29.

44 M. N. Rutkevich, "Razvitie, progress i zakony dialektiki" [Development, Progress and the Laws of the Dialectic], *Voprosy filosofii*, XIX, 8 (August 1965), 26.

45 Konstantinov and Kelle, cited in note 31, p. 12.

46 See Fedoseev's article cited in note 43, especially page 22.

47 V. N. Cherkovets, "O soznatel'nom ispol'zovanii ekonomicheskikh zakonov v sotsialisticheskom obshchestve" [Concerning the Conscious Utilization of Economic Laws in Socialist Society], *Voprosy filosofii*, XIX, 7 (July 1964), 3–13.

48 V. V. Mshvenieradze, "Marksizm i problema tsennostei" [Marxism and the Problem of Values], *Filosofskie nauki*, VIII, 1 (1965), 69.

49 Balakina, cited in note 42, p. 155.

50 For a further development of this topic see my paper, "Value Theory in Soviet Philosophy: A Western Confrontation," *The Proceedings of the XIII International Congress of Philosophy* (Mexico City: Universidad Nacional Autónoma de Mexico, 1963, IV, 133–143. See also S. I. Popov, "Kategorii tsennosti i otsenki i marksistskaia filosofiia" [The Categories of Value and Evaluation and Marxist Philosophy], *Filosofskie nauki*, VIII, 5 (1965), 55–63; A. F. Shishkin, "Chelovek kak vysshaia tsennost'" [Man Considered as Highest Value], *Voprosy filosofii*, XIX, 1 (January 1965), 3–15.

51 For a complete discussion of the development of Marxism-Leninism in Poland, see Z. A. Jordan, *Philosophy and Ideology: The Development of Philosophy and Marxism-Leninism in Poland Since the Second World War* (Dordrecht: Reidel, 1963).

NOTES: CYBERNETICS

1 The first chapter of a Soviet pamphlet on information and control theory was entitled "Ways to Overcome Complexity." See A. I. Berg and Iu. I. Cherniak, *Informatsiia i upravlenie* [Information and Control] (Moscow: "Ekonomika," 1966), pp. 6–22. The authors maintained that the complexity of the national economy in recent years experienced a qualitative leap, but believed that cybernetics provides ways to cope with these new intricacies.

V. M. Glushkov, A. A. Dorodnitsyn, and N. Fedorenko wrote that the application of cybernetics to economic planning would "produce a tremendous national economic effect and at least double the rate of development of the national economy." See their "O nekotorykh problemakh kibernetiki" [On Several Problems of Cybernetics], *Izvestiia*, September 6, 1964, p. 4; English translation in CDSP, XVI, 36 (September 30, 1964), 27–29.

2 A surprisingly large number of the important Soviet publications on cybernetics have been translated into English by the Joint Publications Research Service of the U.S. Department of Commerce. However, the quality of translation is very poor. Two bibliographies, already a bit out-of-date, of both Soviet publications and translations, are David D. Comey, "Soviet Publications on Cybernetics," and L. R. Kerschner, "Western Translations of Soviet Publications on Cybernetics," both in *Studies in Soviet Thought*, IV, 2 (February 1004), 142–177.

3 A similar view is also expressed frequently by philosophers. Ernst Kolman, for example, commented, "The goal of our development—a communist society—is, from the cybernetic point of view, an open, dynamic system with ideal autoregulation." G. S. Gurgenidze and A. P. Ogurtsov, "Aktual'nye problemy dialektiki" [Urgent Problems of the Dialectic], *Voprosy filosofii*, XIX, 10 (October 1965), 147.

4 For a discussion of some of the applications of cybernetics in the economy which Soviet scholars have described, see my "Cybernetics in the Soviet Union," in Walter Z. Laqueur and Leopold Labedz, eds., *The State of Soviet Science* (Cambridge, Mass.: M.I.T. Press, 1965), pp. 6–11.

5 See *Programma kommunisticheskoi partii Sovetskogo Soiuza* [Program of the Communist Party of the Soviet Union] (Moscow: "Pravda," 1961), 71–73.

6 Roger Nett and Stanley A. Hetzler, *An Introduction to Electronic Data Processing* (New York: The Free Press, 1963), pp. 172–173. I am indebted to Mrs. Helen Powers, a student at Columbia University, for pointing out this reference to me.

7 These articles were: V. P. Tugarinov and L. E. Maistrov, "Protiv idealizma v matematicheskoi logike" [Against Idealism in Mathematical Logic], *Voprosy filosofii*, IV, 3 (March 1950), 331–339; M. Iaroshevskii, "Kibernetika—'nauka' mrakobesov" [Cybernetics —A Science of Obscurantists], *Literaturnaia gazeta*, April 5, 1952, p. 4; "Materialist" [pseud.], "Komu sluzhit kibernetika?" [Whom Does Cybernetics Serve?], *Voprosy filosofii*, VII, 5 (May 1953), 210–219. The first of these three attacked cybernetics only indirectly. Several Soviet authors later tried to thrust the blame for the Soviet rejection of cybernetics on "several ignorant bourgeois journalists [who] promoted publicity and cheap speculation about cybernetics." S. L. Sobolev, A. I. Kitov, A. A. Liapunov, "Osnovnye cherty kibernetiki" [The Basic Features of Cybernetics], *Voprosy filosofii*, IX, 4 (April 1955), 136–148.

8 Iaroshevskii, cited in note 7, p. 4.

9 E. Kolman, "Sovremennaia fizika v poiskakh dal'neishei fundamental'noi teorii" [Contemporary Physics in Search of a Future

Fundamental Theory], *Voprosy filosofii*, XIX, 2 (February 1965), 122.

10 E. Kolman, "Chto takoe kibernetika" [What Is Cybernetics?], *Voprosy filosofii*, IX, 4 (April 1955), 148–159.

11 E. Kolman, "Chuvstvo mery" [A Sense of Proportion], in A. I. Berg and E. Kolman, eds., *Vozmozhnoe i nevozmozhnoe v kibernetike* [Possibilities and Impossibilities in Cybernetics] (Moscow: "Nauka," 1964), pp. 51–64. Also, in the same spirit, see Kolman's "Kibernetika stavit voprosy" [Cybernetics Raises Questions], *Nauka i zhizn'*, XXVIII, 5 (May 1961), 43–45.

12 E. Kolman, "O filosofskikh i sotsial'nykh problemakh kibernetiki" [On the Philosophic and Social Problems of Cybernetics], in V. A. Il'in, V. N. Kolbanovskii, and E. Kolman, eds., *Filosofskie voprosy kibernetiki* [Philosophic Problems of Cybernetics] (Moscow: Izdatel'stvo sotsial'no-ekonomicheskoi literatury, 1961), p. 90. In 1963, Kolmogorov wrote that it is theoretically possible for cybernetic devices to experience all activities of man, including emotion. See his "Avtomaty i zhizn'" [Automatons and Life], in Berg and Kolman, eds., cited in note 11, pp. 10–29.

13 Ashby at least once advanced these views to Soviet philosophers in person: "Professor U. Ross Eshbi v redaktsii nashego zhurnala" [Professor W. Ross Ashby in the Editorial Office of our Journal], *Voprosy filosofii*, XIX, 1 (January 1965), 154–157.

14 Certain Soviet writers applied cybernetics to Marxist social theory. L. A. Petrushenko, for example, discussed productive labor as a series of processes performed on the basis of feedback. L. A. Petrushenko, "Filosofskoe znachenie poniatiia 'obratnaia sviaz' v kibernetike" [The Philosophic Significance of the Concept of "Feedback" in Cybernetics], *Vestnik Leningradskogo Universiteta: Seriia ekonomiki, filosofii, i prava*, 17 (1960), pp. 76–86. Akhlibinskii and Khralenko discussed the succeeding stages of history according to Marxism as epochs containing progressively smaller amounts of entropy. B. V. Akhlibinskii and N. I. Khralenko, *Chudo nashego vremeni: kibernetika i problemy razvitiia* [Miracle of Our Time: Cybernetics and Problems of Development] (Leningrad: "Lenizdat," 1963). Kolbanovskii criticized such extensions of cybernetics in his "O nekotorykh spornykh voprosakh kibernetiki" [On Certain Controversial Questions of Cybernetics], in Il'in, *et al.*, eds., cited in note 12, pp. 227–261. But even he referred to the Marxist "withering away of the state" as a cybernetic phenomenon, a moment when society becomes fully self-regulating (p. 248).

15 S. M. Shaliutin, in Il'in, *et al.*, eds., cited in note 12, pp. 25–27.

16 A. I. Berg, "Nauka velichaishikh vozmozhnostei" [A Science of the Greatest Possibilities], *Priroda* (Kazan), LI, 7 (July 1962), 21. Todor Pavlov, honorary president of the Bulgarian

Academy of Sciences, has often written in Soviet publications about his conviction that machines cannot in principle possess emotions. See, for example, his "Avtomaty, zhizn', soznanie" [Automatons, Life, Consciousness], in *Filosofskie nauki*, VI, 1 (1963), 49–57.

N. S. Bud'ko took a devious way out of the riddle about thinking machines by maintaining that although machines can not think, computers are not machines. See his "Iavliaiutsia li mashinami ustroistva, pororabatyvaiushchie informatsiiu?" [Can We Consider Equipment Which Processes Information to be a Machine?] *Voprosy filosofii*, XX, 11 (December, 1966), 75–80.

17 Several writers have maintained that the possibility of thinking machines is ruled out by one principle of dialectical materialism. That principle states that matter at different levels of development possesses qualitative differences. To reduce all complex forms of the movement of matter to combinations of simple forms would mean subscribing, they say, to vulgar rather than dialectical materialism. M. N. Andriushchenko, "Otvet tovarishcham V. B. Borshchevu, V. V. Il'inu, F. Z. Rokhline" [An Answer to Comrades V. B. Borshchev, V. V. Il'in, F. Z. Rokhlin], *Filosofskie nauki*, III, 4 (1960), 108–110. It is clear, however, that the basic "law" underlying this argument—that of the transition of quantity into quality—could be used in favor of the notion of thinking machines. A sufficiently sophisticated arrangement of computer components and perhaps even the integration of organic material—as certain Soviet scientists have suggested—could include a change in "qualitative" relationships.

18 Kolman, "Chuvstvo mery," cited in note 11, p. 53.

19 N. P. Antonov and A. N. Kochergin, "Priroda myshleniia i problema ego modelirovaniia" [The Nature of Thought and the Problem of Its Modeling], *Filosofskie nauki*, VI, (1963), 42.

20 "Norbert Viner v redaktsii nashego zhurnala" [Norbert Wiener in the Editorial Office of Our Journal], *Voprosy filosofii*, XIX, 9 (September 1960), 164.

21 S. L. Sobolev, "Da, eto vpolne seriezno!" [Yes, This Is Completely Serious!], in Berg and Kolman, eds., cited in note 11, pp. 82–88. The philologist's complaint was B. Bialik, "Tovarishchi, vy eto seriezno?" [Comrades, Are You Serious?], *op. cit.*, pp. 77–82.

22 The two-plane solution was a product of several important studies which appeared in 1961 and 1962: A. I. Berg, *et al.*, eds., *Kibernetiku–na sluzhbu kommunizmu* [Cybernetics in the Service of Communism] (Moscow-Leningrad: "Gosenergoizdat," 1961).

In the last few years, several scholars have suggested that information can not be entirely explained on the basis of probability. A. D. Ursul has recently maintained that information must be approached not only probabilistically, but also from other viewpoints, such as topology. This view would indicate that a general theory of matter from the standpoint of cybernetics is still, at best, far in the future. See Ursul's "Nestaticheskie podkhody v teorii informatsii" [Non-statistical Approaches to the Theory of Information], *Voprosy filosofii*, XXI, 2 (February, 1967), 88–97.

23 The crux of the matter here is that thermodynamic entropy is described by an equation, $S = K \log W$, which is functionally the same as Claude Shannon's and Warren Weaver's formula for information: $H = -K \sum_{i=1}^{n} p_i \log p_i$. (In the first equation, S equals entropy of the system, W is the thermodynamic probability of the state of the system, and K equals Boltzmann's constant; in the second equation, H is the amount of information in a system with a choice of messages with probabilities $[p_1, p_2 \ldots p_n]$, and K is a measurement-unit constant.) This seemingly narrow and unimportant coincidence is actually quite exciting and significant; if one could demonstrate that the relationship between neg-entropy and information is an identity, the construction of a general theory of matter covering all complex systems—including human—would seem conceivable. Such a demonstration, is, however, far from being accomplished. Claude Shannon and Warren Weaver, *The Mathematical Theory of Communication* (Urbana: University of Illinois Press, 1949), pp. 19, 105. For a Soviet discussion of this relationship, see I. B. Novik, *Kibernetika, filosofskie i sotsiologicheskie problemy* [Cybernetics, Philosophical and Sociological Problems] (Moscow: "Gospolitizdat," 1963).

24 E. A. Sedov, "K voprosu o sootnoshenii entropii informatsionnykh protsessov i fizicheskoi entropii" [On the Question of the Correlation of the Entropy of Information Processes and Physical Entropy], *Voprosy filosofii*, XIX, 1 (January 1965), 135–145. See also A. D. Ursul, "O prirode informatsii" [On the Nature of Information], *Voprosy filosofii*, XIX, 3 (March 1965), 131–140.

25 I. I. Novinskii, "Poniatie sviazi v dialekticheskom materializme i voprosy biologii" [The Concept of Communication in Dialectical Materialism and Problems of Biology], unpublished dissertation for the degree of doctor of philosophical sciences, Moscow State University, 1963, pp. 324–326. Novik, Biriukov, and Tiukhtin have also attempted to apply the vocabulary of cybernetics to the laws of the dialectic. I. B. Novik, "Kibernetika i razvitie sovremennogo nauchnogo poznaniia" [Cybernetics and

the Development of Contemporary Scientific Knowledge], *Priroda*, LII, 10 (October 1963), 3–11; B. V. Biriukov and V. S. Tiukhin, "Filosofskie voprosy kibernetiki" [Philosophic Problems of Cybernetics], in A. I. Berg, *et al.*, eds., *Kibernetika, myshlenie, zhizn'* [Cybernetics, Thought, and Life] (Moscow: "Mysl'," 1964), pp. 76–108.

26 Compare B. S. Griaznov's comment that cybernetics "does not study phenomena in which there are no processes of control or transfer of information—that is, proooooo of inorganic nature" to Ursul's affirmation that the "law of the accumulation of information holds not only for biological development but also indicates the direction of progress of all matter." B. S. Griaznov, "Kibernetika v svete filosofii" [Cybernetics in the Light of Philosophy], *Voprosy filosofii*, XIX, 3 (March 1965), 162; and A. D. Ursul, cited in note 24, p. 137.

27 A brief discussion of some of these deficiencies is in Rajko Tomovic, "Limitations of Cybernetics," *Cybernetica* (Namur, Belgium), II, 3 (1959), 195–198.

28 G. T. Guilbaud, *What is Cybernetics?* (New York: Grove Press, 1960), p. 3.

29 An interesting discussion of the way in which this confusion has been transferred from certain Western texts to Soviet writings is in Donald P. Bakker, "The Philosophical Debate in the U.S.S.R. on the Nature of 'Information'" (unpublished M.A. thesis, Columbia University, 1966).

30 Stafford Beer, "The Irrelevance of Automation," *Cybernetica*, I, 4 (1958), 288.

NOTES: ECONOMICS

1 The beginning of the discussions of the use of mathematical methods and computers in economics can be dated, as can so many other developments, from Khrushchev's anti-Stalin speech at the Twentieth Party Congress in 1956. The first positive official mention of these methods was in an article by the head of the Central Statistical Administration describing the work of the Congress. See article by V. N. Starovski in *Vestnik statistiki*, 1956, No. 2. For an extensive bibliography and discussion of these developments, see the collection of papers given at the Rochester Conference of May 1965: John Hardt, *et al.*, eds., *Mathematics and Computers in Soviet Economic Planning* (New Haven: Yale University Press, 1967). This volume is hereinafter cited as *Mathematics and Computers*.

2 I will use the term "mathematical economics" in its general

sense. I apply it to the use of mathematical methods to solve economic problems, either theoretical or applied.

3 The essential difference between the Soviet economic system and a capitalist one is the following: in the Soviet system the competition for productive resources (which are always scarce in relation to the desires for output goods and services) takes place within the planning and administrative bureaucracy; the allocation of resources is stipulated in a plan through a series of commands; in turn, these commands are implemented with the aid of a system of positive and negative incentives. Thus, in the allocation of resources in the Soviet economy, prices play a relatively minor role, being used primarily for administrative rather than for allocative purposes. In a capitalist system, the competition for scarce resources takes place within markets, and resources are allocated as a result of the interaction of buyers and sellers in these markets, each pursuing his own interests. Prices play a key role in this process: they indicate to buyers the alternative combinations of goods and services which can be bought with different incomes, and they act as guideposts to producers in the latter's decisions on what and how much to produce, and how to combine different resources in this production. In the market system, prices (and changes in prices) convey to producers information about consumers' desires, and convey to consumers information about the economy's production possibilities.

4 This is not a routine confirmation. Many alterations are apparently made, not least because the plan is usually a very tight one and the ministers try to reduce the pressure that is going to be put on them.

5 Academician A. A. Dorodnitsyn in *Izvestiia*, May 15, 1963, p. 3; cited and translated in an article by L. Smolinski and P. Wiles, "The Soviet Planning Pendulum," *Problems of Communism*, XII, 6 (November–December 1963), 21.

6 V. M. Glushkov in *Literaturnaia gazeta*, September 25, 1962, p. 1. A similar situation in regard to modernization of telephone technology is illustrated by the following quote from *The New York Times*, June 12, 1966, section 4, page 4. "An economist estimated recently that if the telephone system in the United States tried to operate with today's volume of calls but the equipment of the 1930's, it would take every able-bodied female in the nation between the ages of 14 and 65 to run the switchboards."

7 For surveys of this literature, see the paper by Richard Judy in *Mathematics and Computers;* Smolinski and Wiles, *op. cit.;* and Leon Smolinski, "What Next in Soviet Planning?" *Foreign Affairs*, XLII, 4 (July 1964), 602–613.

8 Judy, in *Mathematics and Computers*, pp. 17–32.
9 *Ibid*, pp. 31–32. One Soviet economist has stated that only 10 per cent of all the data collected are ever used for planning and management purposes. N. Fedorenko, "O rabote tsentralnogo ekonomikomatematicheskogo instituta" [On the Work of the Central Institute of Mathematical Economics], in *Vestnik Akademii Nauk*, 1964, No. 10, p. 6, cited in Judy, *Mathematics and Computers*, p. 32.
10 See, for example, V. M. Glushkov and N. Fedorenko, "Problemy shirokogo vnedreniia vychislitelnoi tekhniki v narodnoe khoziaistvo" [Problems of Widespread Introduction of Computer Technique in the National Economy], *Voprosy ekonomiki* 1964, No. 7, pp. 87–92; V. Pugachev, "Voprosy optimal'nogo planirovaniia narodnogo khoziaistva s pomoshch'iu edinoi gosudarstvennoi seti vychislitel'nikh tsentrov" [Questions of Optimum National Economic Planning with the Aid of a Single State Network of Computer Centers], *ibid.*, pp. 101–103.
11 N. Gal'perin, "Sovershenstvovaniye material'no-tekhnicheskogo snabzheniya i borba protiv mestnicheskikh tendentsii" [Improvement of Material-Technical Supplies and the Battle Against Localizing Tendencies], *Voprosy ekonomiki*, 1958, No. 7, p. 46.
12 For Western discussions of Soviet input-output, see the paper by Vladimir Treml in *Mathematics and Computers*; and, Herbert S. Levine, "Input-Output Analysis and Soviet Planning," *American Economic Review*, LII, 2 (May 1962), 127–137.
13 V. Belkin, "O primenenii elektronnikh vychislitel'nikh mashin v planirovanii i statistike narodnogo khoziaistva" [Concerning the Application of Electronic Computers in Planning and Statistics for the National Economy], *Voprosy ekonomiki*, 1957, No. 12, pp. 139–147; V. Nemchinov, "O sootnosheniiakh rasshirennogo vosproizvodstva" [On Relationships in an Expanding Economy], *Voprosy ekonomiki*, 1958, No. 10, pp. 20–31; and "Balansovyi metod v ekonomicheskoi statistike" [The Balance Method in Economic Statistics], *Uchenye zapiski po statistike* [Learned Papers in Statistics] (Moscow: U.S.S.R. Academy of Sciences, 1959), Vol. 4, pp. 5–20.
14 L. Kantorovich, *Matematicheskie metody organizatsii i planirovaniia proizvodstva* [Mathematical Methods of Organizing and Planning Production] (Leningrad: Leningrad University Press, 1939).
15 L. Kantorovich, *Ekonomicheskii raschet nailuchshego ispol'zovaniia resursov* [The Economic Calculation of the Best Use of Resources] (Moscow: U.S.S.R. Academy of Sciences, 1959). English translation as *The Best Use of Economic Resources* (Cambridge, Mass.: Harvard University Press, 1965).

16 For a survey of this literature, see the papers of Benjamin Ward and John M. Montias in *Mathematics and Computers.*

17 For an interesting discussion, see Robert W. Campbell, "Marx, Kantorovich and Novozhilov, Stoimost' versus Reality," *Slavic Review*, XX, 3 (September 1961), 402–418.

18 A. Vainshtein, "Nyet, eto ne moda" [No, It's Not the Fashion], *Literaturnaia gazeta*, July 23, 1964, p. 2.

19 See *Mathematics and Computers*; also the article, "Nauka chelovecheskogo shchastia" [The Science of Human Happiness], by N. Fedorenko, the director of the Central Institute of Mathematical Economics, U.S.S.R. Academy of Sciences, in *Literaturnaia gazeta*, March 29, 1966, p. 2 (English translation in JPRS 35, 486).

20 See, for example, *Literaturnaia gazeta*, May 14, 1964, p. 2. This was the first article in a running debate on the use of mathematics in economics which appeared in *Literaturnaia gazeta* on May 30, June 18, 25, July 23, August 6, 27, 1964. The articles have been compiled and translated in the Joint Publications Research Service (U.S. Department of Commerce), No. 27,069 (October 26, 1964). Moreover, some concern has been expressed that in the great quantity of work and writing being done in this field, quality has often suffered. In an article that has received wide attention since it appeared in *Izvestiia*, four leading mathematicians and economists (Dorodnitsyn, Kolmogorov, Gnedenko, Vainshtein) scathingly criticized some recent books on the use of mathematical methods in economics, including a book by the director of the computer center at Gosplan, N. N. Kovalev (*Izvestiia*, January 24, 1965, p. 2). January 1965 was a bad month for Kovalev, for an article of his was also criticized in an open letter by nine specialists, in *Voprosy ekonomiki*, 1965, No. 1, pp. 153–155. Another worried economist speaks about the development of a "fad," especially in regard to linear programing, and its negative effect on quality. See Ia. Gerchuk, *Granitsy primeneniia lineinogo programmirovaniia* [The Limits on the Use of Linear Programing] (Moscow: "Ekonomika," 1965), p. 5.

21 See for example, Ia. Gerchuk, *op. cit.*, p. 11; and the article by Vainshtein cited in note 18.

22 "Avtomatizatsiia i upravlenie ekonomikoi" [Automation and Economic Management], by G. Kazanski, *Izvestiia*, April 26, 1966, p. 3.

23 On this, see the paper by Treml in *Mathematics and Computers.*

24 An 83-order value table was involved. See V. Treml, "Economic Interrelations in the Soviet Union," in U.S. Congress, Joint Economic Committee, *Annual Economic Indicators for the*

U.S.S.R. (Washington, D.C.: Government Printing Office, 1964), pp. 185–213.

25 Again the economist made use of both physical units (346 products) and value terms (83 sectors). An informative discussion of the value table is to be found in L. Berri, F. Klotsvog, S. Shatalin, "Opyt rascheta eksperimental'nogo planovogo balansa za 1962 god" [The Experience of the Construction of the Experimental Planning Balance for 1962], *Planovoe khoziaistvo,* 1962, No. 9, pp. 34–43.

26 For some discussion of the table in physical units, see N. N. Kovalev, "Nekotorye problemy postroeniia mezhotraslevykh balansov v naturalnom vyrazhenii" [Some Problems of the Construction of Interindustrial Balances in Physical Terms], *Voprosy ekonomiki,* 1963, No. 5, pp. 76–88.

27 A. Klinski, reporting on a conference paper by Efimov, Shatalin, and Klotsvog, *Planovoe khoziaistvo,* 1964, No. 7, p. 91.

28 *Ibid.*

29 See Fedorenko's article, cited in note 19, p. 2.

30 Note the tone of optimism expressed at the 1960 conference on mathematical methods in the coordinating plan and schedule for scientific work and the introduction of these methods into practice (prepared by the Computing Center of the U.S.S.R. Academy of Sciences). See the article by V. Belkin, "O plane koordinatsii rabot po primeneniu matematicheskikh raschetov" [Concerning the Plan for Coordination of Work Through Application of Mathematical Methods and Electronic Computers in Economic Calculations] in V. Nemchinov, ed., *Obshchie voprosy primeneniia matematiki v ekonomike i planirovanii* [General Problems of the Use of Mathematics in Economics and Planning] (Moscow: U.S.S.R. Academy of Sciences, 1961), pp. 129–149.

31 See Fedorenko's article cited in note 19, especially p. 2.

32 V. Cherniavskii, "Kriterii optimalnosti" [The Criteria of Optimality], *Ekonomicheskaia gazeta,* March 17, 1965, p. 9.

33 See the paper by Judy in *Mathematics and Computers,* and V. M. Glushkov, "Vychislitel'nuiu tekhniku v upravlenie narodnym khoziaistvom" [Put Computer Techniques into the Management of the National Economy], *Pravda,* July 12, 1964, p. 4.

34 N. Fedorenko, "Vychislitel'nuiu tekhniku v narodnoe khoziaistvo" [Put Computer Techniques into the National Economy], *Ekonomicheskaia gazeta,* October 27, 1965, p. 18.

35 See, for example, A. Aganbegian, quoted in an article, "Matematizatsiia ekonomiki—Da ili nyet?" [The Mathematization of Economics—Yes or No?] by A. Smirnov-Cherkezov, *Literaturnaia gazeta,* May 14, 1964, p. 2; and the editorial article in *Ekono-*

mika i matematicheskie metody [Economics and Mathematical Methods], November-December 1965, p. 820.

36 See the account of the report by M. Eidelman at the 1961 conference on the use of mathematical methods in economics, in *Voprosy ekonomiki*, 1962, No. 1, pp. 117–118.

37 See Treml in *Mathematics and Computers*, pp. 75, 109; and M. Eidelman, "Metodologicheskie problemi otchetnogo mezhotraslevogo balansa" [Methodological Problems of Accounting for Interindustrial Balance], *Vestnik statistiki*, 1963, No. 5, p. 17.

This, of course, did not eliminate the particularly vexing problem of aggregation common to all input-output work. Whenever the different products grouped together in a sector are produced by different input combinations, the input coefficient for the sector (a weighted average of the input coefficients of the individual products within the sector) has a tendency to be unstable. It will change if changes in the output level of the sector (which may occur in the process of plan construction) cause the shares of different products in the sector to change.

Another problem inherent in the input-output method is its "linearization" of relationships. In practice, of course, the relationship of inputs to a unit of output is not constant over all levels of output. And much of the essence of economic growth lies in the economies (and diseconomies) that are associated with increasing levels of output. While this may not be an important drawback of the method in short-term planning (the material-balances method is similarly deficient in this respect because it is also substantially based on linear relations between inputs and outputs), it would be significant in long-term planning.

In Western uses of input-output method, it is the instability of input coefficients resulting from technological change that has caused much concern. This need not be so difficult a problem for Soviet users of input-output in short-term plan construction, because technological change can, at least so it is claimed, be foreseen and planned, and the necessary adjustments made in the coefficients. In long-term planning, however, this would be a serious drawback of the method.

38 This "pure sector–administrative target" problem may be one argument in favor of using input-output for the construction of five-year plans, which are nonoperational plans in the sense that direct commands are not given to any organization on the basis of the targets in the plan.

39 Western computer language has an expression for this relationship between bad data and useless results: GIGO—"garbage in, garbage out." In Western use of input-output, empirically observed relationships between inputs and outputs (input norms)

are assumed to be rational because they are the result of decision-making within the context of the market. In Soviet use of input-output, however, empirical input norms merely reflect past planning decisions and thus have no claim to economic rationality. For a discussion of this point, see *Voprosy ekonomiki*, 1966, No. 6, p. 158.

40 It was said that for the 83-sector 1962 planning table in value terms there were 4,260 direct input norms (62 per cent of the cells were occupied). See the article by Berri, Klotsvog, and Shatalin, cited in note 25, *Planovoe khoziaistvo*, 1962, No. 9, p. 38. The number of direct input norms in the 346-sector 1962 planning table in physical terms was about 10,000 (Kovalev, cited in note 26, p. 80).

41 In fact, one Soviet economist recently argued that computers used with old methods often produce worse results than those obtained without computers. See N. Fedorenko's article, cited in note 19, p. 2.

42 V. Belkin and I. Birman, "Samostoiatel'nost' predpriiatiia i ekonomicheskie stimuli" [Independence of the Enterprise and Economic Stimuli], *Izvestiia*, December 4, 1964, p. 5; English translation in CDSP, XVI, 50 (January 6, 1965), 14–15.

43 This I have elsewhere been at pains to point out: *Soviet Studies*, XV, 3 (January 1965), 352–356.

44 V. Nemchinov, ed., *Primenenie matematiki v ekonomicheskikh issledovaniiakh* [The Use of Mathematics in Economic Research] (Moscow: Izdatel'stvo sotsial'no-ekonomicheskoi literatura, 1959), p. 7.

45 *Ekonomicheskaia gazeta*, March 10, 1965, p. 10. See also M. Golanskii, "Stoimost' i raschetnye tseny" [Value and Accounting Prices], in *Planirovanie i ekonomiko-matematicheskie metody* [Planning and Mathematical Economic Methods] (Moscow: "Nauka," 1964), especially pp. 389–392. This article appears in English in the journal *Mathematical Studies in Economics and Statistics*, I, 3 (Spring 1965). Cf. Golanskii with Campbell, in *Slavic Review*, XX, 3 (September 1961), 402–418.

46 See Fischer's chapter in this volume.

47 This is the function price plays in the system of "khozraschet," or economic accountability.

48 See the interesting article, "O kriterii optimal'nosti narodnokhoziastvennogo plana" [Concerning the Criteria of Optimality of the National Economic Plan], by B. Smekhov, *Voprosy ekonomiki*, 1965, No. 1, especially pp. 124–127. In his 1959 book, Kantorovich skirted some basic difficulties by assuming an objective function in which all final demands were combined in fixed proportion to one another. He then proceeded to maximize the output level of this "composite product."

49 I. Birman, "Dreif v tikhoi zavodi" [Adrift in Still Waters], *Litera-turnaia gazeta*, June 18, 1964, p. 2, says that Stalin had pro-claimed a criterion of optimality (the so-called principle of "higher profitability") but that this could be used to justify "everything on earth."

50 John M. Montias, "Moscow University Conference on Mathe-matical Economics of March 1962," *ASTE Bulletin* (Fall 1963), p. 10.

51 Likewise, are mathematical economists and social scientists in general wholly committed to "pure" objectives? In fact, it is naive to assume that they will balk at applying their skills toward the accomplishment of "impure" objectives, or that they will necessarily concern themselves with the morality of objec-tives. See "MSU: The University on the Make," *Ramparts*, IV, 12 (April 1966), 11–23. Moreover, what is involved here is not really a question of morality but a policy question concern-ing the shares of various components of the national product.

52 There were some rare exceptions. See the article by L. Dudkin "Matematiko-ekonomicheskaia skhema optimal'nogo material'-nogo balansa sotsialisticheskogo narodnogo khoziaistva" [Math-ematical-Economic Scheme of the Optimum Material Balance in the Socialist National Economy], in *Problemy optimal'nogo planirovaniia: Proektirovaniie i upravleniie proizvodstvom* [Prob-lems of Optimal Planning: Designing and Management of Pro-duction] (Moscow: Moscow University Press, 1963).

53 See the article by Glushkov and Fedorenko, cited in note 10, p. 90.

54 V. Dadaian, "Printsip optimuma v narodnom khoziaistve" [The Principle of the Optimum in the National Economy], *Ekono-micheskaia gazeta*, September 1, 1965, p. 6.

55 See V. Dadaian, "Optimal'noe rukovodstvo narodnym khoziai-stvom i zadachi ekonomicheskoi kibernetiki" [The Optimum Leadership of the Economy and the Tasks of Economic Cy-bernetics], *Vestnik Moskovskogo Universiteta: seriia VII, Eko-nomika* (January–February 1966), pp. 52–59. See also Fedo-renko's article cited in note 19, p. 2, and I. Ivanov, "Primenenie mezhotraslevogo balansa v planirovanii" [The Application of the Interindustry Balance in Planning], *Planovoe khoziaistvo*, 1965, No. 11, p. 24.

56 See Abraham S. Becker in *Mathematics and Computers*.

57 However, see V. Pugachev's article cited in note 10, and the discussion of his model in almost all the papers in *Mathematics and Computers*.

58 "From the standpoint of methodology, economics can be regard-ed as a very complicated cybernetic system, the elements of which change in time. This approach permits the application,

on the basis of the laws of Marxist-Leninist political economy, of precise quantitative methods during the investigation and modeling of economic processes. In general, it must be considered that such a complex system as the economy of the U.S.S.R. cannot be completely determined and its elements cannot be fully described quantitatively. If one breaks the economy down by levels—enterprise, association, branch, republic, national economy as a whole—it will be found that the lower the level, the stronger is the effect of random factors and the greater is the extent of the spontaneous nature of management; or as the specialists say, stochastic unprognosticated. This also makes it necessary to establish a multi-stepped and sufficiently decentralized system, the individual elements of which strive toward the realization of a national economic optimum through the use of principles of self-adjustment." N. Fedorenko, article cited in note 19, p. 2.

59 V. Nemchinov, "Sotsialisticheskoe khoziaistvo i planirovanie proizvodstva" [The Socialist Economy and the Planning of Production], *Kommunist*, 5 (March 1964), 74–88.

60 See V. Dadaian, cited in note 55, pp. 52–59.

61 On the role of ideology in economic development, see Alexander Gerschenkron, *Economic Backwardness in Historical Perspective* (Cambridge, Mass.: Harvard University Press, 1962), pp. 22–26.

INDEX OF NAMES

The Index was compiled by Wesley A. Fisher and Stephen T. Kerr.

Printed in Great Britain
by Amazon

29888201R00106